Telomerase and non-Telomerase Mechanisms of Telomere Maintenance

Edited by Tammy A. Morrish

Published in London, United Kingdom

IntechOpen

Supporting open minds since 2005

Telomerase and non-Telomerase Mechanisms of Telomere Maintenance
http://dx.doi.org/10.5772/intechopen.73734
Edited by Tammy A. Morrish

Contributors
Snehasish Nag, Prem Singh Yadav, Abubakar Muhammad Wakil, Raymund J. Wellinger, Yulia Vasianovich, Alexandra Krallis, Marie Jirkovská, Marie Korabečná, Soňa Laššáková, Julien Audry, Kurt Runge

First published in London, United Kingdom, 2020 by IntechOpen
IntechOpen is the global imprint of INTECHOPEN LIMITED, registered in England and Wales, registration number: 11086078, 7th floor, 10 Lower Thames Street, London, EC3R 6AF, United Kingdom
Printed in Croatia

British Library Cataloguing-in-Publication Data
A catalogue record for this book is available from the British Library

Additional hard and PDF copies can be obtained from orders@intechopen.com

Telomerase and non-Telomerase Mechanisms of Telomere Maintenance
Edited by Tammy A. Morrish
p. cm.
Print ISBN 978-1-78984-574-7
Online ISBN 978-1-78984-575-4
eBook (PDF) ISBN 978-1-83880-281-3

Meet the editor

Tammy A. Morrish's scientific interests were focused on understanding mechanisms of non-LTR retrotransposition during DNA repair and telomere maintenance in tumors and primary cells. From 2012-2015, she was an Assistant Professor at the Department of Biochemistry and Cancer Biology at the University of Toledo. Her postdoctoral training was at Johns Hopkins University in the Department of Molecular Biology and Genetics where she received the Howard Temin Pathway to Independence Award from the NIH/NCI. Dr Morrish was also an American Society of Hematology Fellow Scholar and received funding from the Leukemia and Lymphoma Society. She received her Ph.D. from the University of Michigan in the Department of Human Genetics and was awarded a Rackham Predoctoral Fellowship and an Overall Excellence in Research and Service Award. She also had a B.A. in biochemistry from the University of Texas.

A note from publisher:

It is with great sadness and regret that we inform the contributing authors and readers of this book that the Editor, Dr Tammy A. Morrish, passed away during the publishing process of the book and before having a chance to see its publication. The book *Telomerase and non-Telomerase Mechanisms of Telomere Maintenance* was her first edited volume with us. Fruitful collaboration continued until her final days. We would like to acknowledge Dr Morrish's contribution to scientific publishing, which she made during years of dedicated work, and express our gratitude for her pleasant cooperation with us.

Contents

Preface

This book is a collection of reviewed and relevant research chapters on the developments within the telomerase and non-telomerase mechanisms of telomere maintenance field of study. The book includes scholarly contributions by various authors and has been edited by a group of experts in the fields of biochemistry, genetics, and molecular biology. Each contribution comes as a separate chapter, complete in itself but directly related to the book's topics and objectives.

The book includes chapters dealing with the topics: telomerase structure and function, activity and its regulation with emerging methods of measurement in eukaryotes, telomerase in space and time: regulation of yeast telomerase function at telomeres and DNA breaks, telomere formation systems in budding and fission yeasts, syndromes associated with telomere shortening and telomeres and telomerase activity in the human placenta.

The target audience comprises scholars and specialists in the field.

Chapter 1

Telomerase Structure and Function, Activity and Its Regulation with Emerging Methods of Measurement in Eukaryotes

Prem Singh Yadav and Abubakar Muhammad Wakil

Abstract

The telomerase reverse transcriptase has an essential role in telomere maintenance which is very important in aging process and cancer biology. Recent studies have revealed three-dimensional architecture of both human and ciliate telomerase at about 25 Å resolution, using single particle electron microscopy (EM). Telomerase supplements the tandem array of simple-sequence repeats at chromosome ends to compensate for the DNA erosion inherent in genome replication which makes it to be distinct among polymerases. Telomeres are found at the end of eukaryotic linear chromosomes and proteins that bind to them and help to protect DNA from being recognized as double-strand breaks thus preventing end-to-end fusions. The activity of telomerase is tightly regulated at multiple levels of cellular development, from transcriptional regulation of the telomerase components to holoenzyme biogenesis and recruitment to the telomere site for activation and processing. Commonly used methods in telomere biology are telomere restriction fragment (TRF), telomere repeat amplification protocol (TRAP) and telomere dysfunction induced foci (TIF) analysis. This chapter summarizes our current knowledge on the mechanisms of telomerase recruitment and activation using insights from studies in mammals and budding and fission yeasts. Finally, we discuss the differences in telomere homeostasis between different cell types and non-telomerase telomere maintenance mechanisms.

Keywords: telomerase, telomere, holoenzyme biogenesis, chromosome ends, reverse transcriptase

1. Introduction

Telomerase is a ribonucleoprotein complex, composed of a reverse transcriptase enzyme catalytic subunit and a long non-coding RNA that contains the template sequence for telomere synthesis [1, 2] and is required for linear chromosome maintenance in most eukaryotes. The enzyme telomerase is active in germ cells and during early embryogenesis, ensuring restoration of telomere length for the next generation. However, when using an aged somatic cell with shortened telomeres for cloning, the offspring might start with a diminished replicating capability of its cells and consequently age, or at least reach senescence, faster [3]. Telomere biology differ significantly among mammalian species, ranging from humans with very short telomeres and limited telomerase activity in the cells to mice with extremely long telomeres and active telomerase in multiple tissues. Human fibroblast cells have been reported to possess short telomeres and suppress telomerase activity [4]. Hence, human cells undergo permanent growth arrest, or replicative senescence, which is triggered by the critically short telomeres upon serial passaging in culture [5, 6]. Replicative senescence is an important barrier in tumor progression, as malignant tumors must reactivate telomerase or use the alternative lengthening of telomere mechanism to gain unlimited proliferation potential [7]. The expression of some components of the telomerase holoenzyme is tightly regulated [8]. For instance, in unicellular eukaryotes TERT and TER are constitutively expressed. In mammals, TERT is expressed only in highly proliferative cells and tumor cells. Somatic cells and cells with low proliferative capacity lack enzyme activity, this is the reason why telomerase activity is extensively studied as a potential target for antitumor therapy [9, 10]. However, apart from telomerase there are other mechanism used to maintain chromosome length; in some organisms, such as the fruit fly *Drosophila melanogaster*, retrotransposon-like elements are alternatively used to replenish the DNA at the ends of chromosomes [11]. As reported by Miriam Aparecida Giardini et al. [12], under certain circumstances, yeast and human cells that lack telomerase activity, as well as some telomerase-negative tumor lineages, are able to maintain their telomeres using a recombination-based DNA replication mechanism known as alternative lengthening of telomeres (ALT) [13, 14].

Research conducted in the past 10 years has revealed important discoveries on the evolution of telomere maintenance mechanisms [15]. Telomeres serve as substrates for telomerase, the enzyme responsible for adding DNA to the ends of chromosomes, thus maintaining chromosome length [9, 16]. To compensate for the DNA erosion inherent in genetic stability, telomerase adds tandem array of simple-sequence repeats at the chromosome ends. The template for telomerase reverse transcriptase is within the RNA subunit of the ribonucleoprotein complex, this contains additional telomerase holoenzyme protein components within cells that assemble the active ribonucleoprotein and promote its function at telomeres. In terms of its reiterative reuse of an internal template, telomerase is different among other polymerases [17]. Like many polymerases, telomerase catalyzes nucleotide addition to a primer 3′ hydroxyl group, forming a product-template duplex. Accordingly, telomerase and other polymerases share a metal-dependent chemistry of nucleotide addition. Beyond these parallels, telomerase possesses unique properties of nucleic acid handling. Accurate telomeric repeat synthesis depends on strict boundaries of template copying within TER. Also, telomerases from most species studied have the exceptional ability to extend a primer by processive addition of repeats (reviewed in [17]). Repeat addition processivity (RAP) obliges dissociation of the product-template duplex without product dissociation from the enzyme [17]. The template-dissociated single-stranded DNA must maintain template-independent interactions while the template repositions for base pairing of its 3′ end, rather than the 5′ end, with the product. These coordinated nucleic acid handling events transpire as part of the full catalytic cycle of repeat synthesis [17].

Telomeres are specialized nucleoprotein structures located at the ends of linear chromosomes; they consist of TTAGGG repetitive sequences. They function to prevent natural chromosomal termini from activating the DNA damage response. [18]. The knowledge and understanding of telomerase structure, mechanism of action and factors involved in its activity would give more insight in overcoming the problem of replicative senescence.

2. Telomerase organizational architecture

During the last 4 years, progress in telomere research has revealed the three-dimensional architecture of telomerase in human and ciliate which is measured at about 25 Å resolution, this was obtained using single particle electron microscopy (EM). The structural analysis of the two holoenzyme complexes isolated from cells revealed that telomerase in ciliate is monomeric while the human telomerase is dimeric and it is only functional as a dimer [19]. Telomerase is a RNP complex with high-molecular weight and comprises of two major components and these are; TERC and a TERT (**Figure 1**) [12]. TERC is the RNA component which is essential for telomere synthesis; this serves as a template to elongate the 3′ overhang of the telomeric G-rich strand and specifies the repeat sequence added. In vertebrates, the TERC is comprised of three highly conserved structural domains and these are: the template pseudoknot domain, CR4-CR5 domain, and the small Cajal-body RNA domain. The template pseudoknot domain contains the template region for telomeric DNA synthesis and a conserved pseudoknot structure crucial for telomerase activity [20, 21]. Among eukaryotes the RNA component varies dramatically in sequence composition and in size [22–26].

TERT contains catalytic domains and is the protein component which acts as a specialized reverse transcriptase. In humans, TERT and TERC are the components required for telomerase activity in *in vitro* condition, although in *in vivo* condition some proteins are associated with the holoenzyme complex and are also essential for the catalytic function of telomerase enzyme [27]. About 32 different proteins are associated with human telomerase *in vivo* so as to maintain its functionality [28]; but few of these proteins are phylogenetically conserved. Proteins associated with telomerase activity have been best categorized in eukaryotes. Ciliate telomerase RNPs complex comprises a telomerase-specific La motif protein that folds telomerase RNA into a conformation that will be recognized by the TERT component [29].

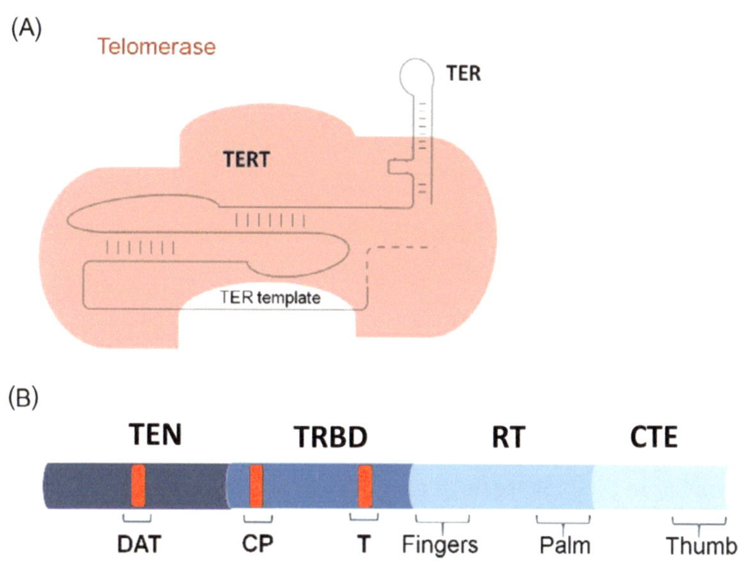

Figure 1.
Telomerase holoenzyme showing the various components. (A) Telomerase reverse transcriptase component (TERT) and telomerase RNA component (TER). (B) Diagram representing the TERT primary structure showing important TERT domains which include; the telomerase N-terminal domain (TEN), telomerase RNA-binding domain (TRBD), reverse transcriptase domain (RT) and C-terminal extension region (CTE). The position of the structural fingers, palm, and thumb subdomains are also highlighted. Openly accessed from Miriam Aparecida Giardini et al. [12] and Nanda Kumar and Cech [34].

At least two additional subunits, p45 and p75 are present in *Tetrahymena thermophila* holoenzyme which are not considered essential for TERT RNP assembly but are required for telomere elongation process. This common role is played by the proteins Est1p and Est3p in *Saccharomyces cerevisiae* [30]. There has not been any report in humans for proteins with this type of function, but known interaction partners of human telomerase RNA (hTR) have been reported and they include dyskerin (the H/ACA-motif RNA binding proteins), NHP2, NOP10, and GAR1. These four proteins assemble with hTR and with large families of H/ACA-motif small nucleolar (sno) RNAs and small Cajal body (sca) RNAs (**Figure 2**). Amino acid substitutions in dyskerin reduce hTR accumulation and this give rise to the X-linked form of dyskeratosis congenita (a bone marrow failure syndrome) [31]. Proteomics of highly purified active human telomerase led to the suggestion that only hTERT and dyskerin are associated with hTR [28]. However, this conclusion is challenged by previous studies showing that dyskerin possesses minimal RNA binding affinity in the absence of its H/ACA-motif binding partners NHP2 and NOP10 [32].

In general, telomerase RNP complexes exhibit conserved compositions and structures, even in evolutionarily distant organisms. Their compositions are similar from yeasts to mammals, including humans (**Figure 1**) [5, 12, 33].

For *in vitro* enzyme activity, minimal complex formation by TERT and TER components is sufficient. Nevertheless, *in vivo*, enzyme biogenesis, enzyme activity, and nucleotide addition processes also depend on other accessory proteins, indicating that a relatively complex maturation pathway is required for generation of an active RNP that has to find its substrate [35, 36]. Telomerase function to avoid the loss of terminal DNA, which is caused due to inability of DNA polymerases to completely replicate the 5′ ends of linear DNA molecules and also the actions of exonucleases involved. Both processes are responsible for generation of transient 3'OH overhangs found on the opposite ends of the leading and lagging DNA strands. The recognition of these overhangs are done by the end-binding proteins, they bind to the overhangs and afterward recruit telomerase to elongate the G-strand termini. The C-strand is synthesized by the conventional DNA replication pathway as soon as the telomeres are replicated by the telomerase [2, 14, 37, 38].

2.1 Telomerase-associated proteins

Though expression of hTERT and hTERC in rabbit reticulocyte lysates is sufficient to reconstitute basic telomerase enzyme [29], but the *in vivo* requirements for other factors necessary in the assembly of the active enzyme which did not clearly revealed some of this *in vitro* reconstitution, even though some of these factors are present in the rabbit reticulocyte lysates [39]. The molecular chaperones; Hsp90 and p23 are present in rabbit reticulocyte lysates. These are directly associated with the hTERT and are necessary for telomerase activity [39]. Biochemical and genetic studies reveal that additional protein subunits of telomerase exists which may be involved in the biogenesis or assembly of active telomerase RNP complex and may facilitate or regulate the access of telomerase to its substrate (i.e. the telomeres) [39].

2.2 hTERT-associated proteins

Biochemical fractionation of telomerase activity from the yeast *Tetrahymena thermophila* was used in identifying the first telomerase-associated proteins [29, 39]. The proteins, p80 and p95, were identified by their association with the RNA

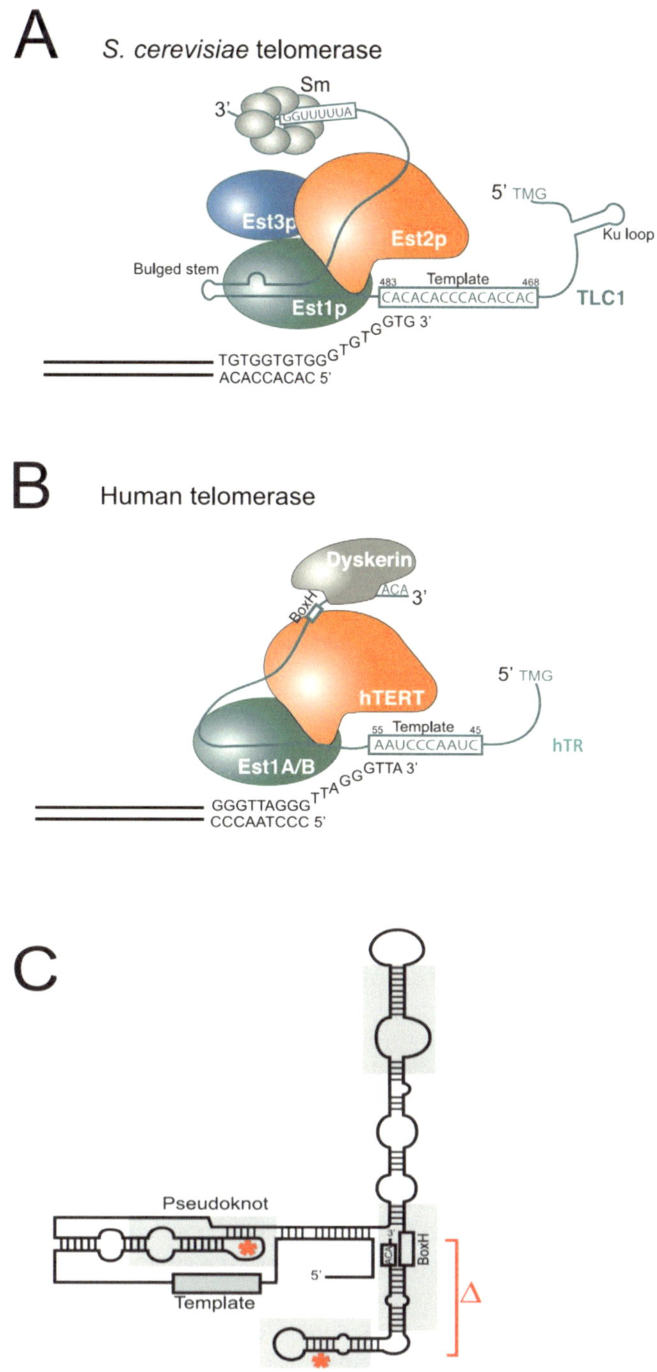

Figure 2.
*Showing schematic structure of (A) budding yeast (*S. cerevisiae*) telomerase, (B) human telomerase cropped at a telomere 3'end, and (C) vertebrate telomerase RNAs showing the conserved structural motifs. The positions of* DKC *mutations in the human telomerase (*hTERC*) gene are shown in red. Images are adopted from Smogorzewska and de Lange [33].*

component of telomerase and by copurification of telomerase activity [37]. In a similar report, *Tetrahymena* strains were shown to lack p80 and p95, and the levels of telomerase activity with its RNA appear to function totally normal. This suggests that these proteins are not core components of telomerase and can be a separate ribonucleoprotein that was copurified nonspecifically with telomerase [40]. Nevertheless, it was reported in another study that cells devoid of p80 and p95 have their telomeres elongated both in macronuclei and micronuclei but lose genetic content in their micronuclei, which suggest the role of p80 and p95 proteins in micronuclear genomic stability and telomere length maintenance [40]. TEP1 (telomerase-associated protein 1) which was identified in humans, mice and rats is the mammalian homolog of p80, and is involved with telomerase activity [39]. TEP1 consists of 2629 amino acids, much larger than p80. About 900 amino acids found at the amino terminus of TEP1, contain region homologous to p80 which were found to associate with telomerase RNA. The carboxyl terminus of TEP1 contains 12 WD40 repeats, a motif known to be involved in protein-protein interactions [39]. TEP1 expression can be distinguished in most tissues irrespective of telomerase activity. Disturbance of mouse TEP1 has no effect on telomerase activity or telomere length in spite of its association with both the RNA and catalytic components of telomerase in cell extracts from immortalized human, mouse, and rat cells [39, 41]. The TEP1 protein has also been recognized as a constituent of large cytoplasmic particles called vaults, which are ribonucleoprotein complexes [39]. The functions of TEP1 in both telomerase and vaults are still not elucidated [39]. The molecular chaperone p23 was first identified to be associated with hTERT using the amino terminus (amino acids 1–195) as the desirability in a yeast two-hybrid screen. Consequently, it was observed that the proteins p23 and p90 were in association with hTERT in mammalian cells and in *in vitro* condition [41]. The first identified sets of proteins which interact physically and functionally with human telomerase is the hsp90 chaperone complex and have been found to support complete assembly of ribonucleoprotein and the formation of active telomerase enzyme [39]. It is well known that other reverse transcriptase that are of viral origin also interact with hsp70, hsp90, and p23, but appear to be transient [12].

3. Telomerase activity in different cell types

Telomeres are needed to maintain the ends of chromosomes and sustain chromosome stability in eukaryotic cells. Telomeres loss their noncoding DNA sequences in the erosion that happens during DNA replication in each cell cycle. They do this to protect the genetic information in the chromosomes [42, 43]. Most somatic cells enter into replicative senescence because they have undergone sufficient cell divisions to cause critical shortening of the telomeres. Some cells, including lymphocytes, germ cells, stem cells and unicellular eukaryotes such as yeast, express the enzyme telomerase, which gives them the ability to replenish their telomeres and give them further replicative potential [30, 44]. Most human tumors express active telomerase enzyme making them immortal while in differentiated cells, expression of the telomerase components is closely regulated [45, 46]. A direct correlation between continuous cell division and telomere length maintenance was studied in *in vitro* culture condition through ectopic expression of telomerase activity in somatic cell [47]. Even though cancer cells steadily maintain telomere length which also tend to be shortened in later stage [48, 49], some of them are critically shortened and are termed 't-stumps' [50] resulting in immortal cells which possessed a high risk of chromosome instability. This is extraordinarily different from our understanding of telomerase activity in normal cells, in which telomerase acts to elongate shorter telomeres until they are no longer short [7, 51]. The reason why telomerase behaves differently in cancer cells still remains an area of interest for research.

3.1 Telomerase expression and cellular proliferation

The expression of telomerase enzyme activity in different types of cells has been characterized using the telomeric repeat amplification protocol (TRAP) assay. The method fundamentally measures the telomerase activity confined within a cell lysate *in vitro* culture [10]. By using this assay, it is well documented that most differentiated somatic cells lack detectable telomerase activity [10, 46], explaining the reason why telomeres shorten in each cell divisions [47, 48, 52]. In adult testes and ovaries, telomerase enzyme is highly expressed thereby, allowing consistently longer telomeres to be inherited by the next generation [48, 53]. During the early embryonic development, telomerase enzyme remains active but its expression declines after the blastocyst stage and cannot be detected in neonatal somatic cells [53–55]. Telomerase activity is weak in most stem cell populations [10, 44, 46, 56], this is not sufficient to immortalize cells but can extend the proliferative capacity of these cells (reviewed in [6, 57]). Remarkably, the Hayflick limit of somatic cells can be indefinitely avoided when telomere length is maintained by high expression of telomerase activity [47, 58]. Therefore, the level of telomerase activity and its expression determines the level of telomere length elongation and proliferative ability of a cell.

4. Regulation of telomerase activity in mammals

Telomerase activity is widely regulated owing to its important role in the maintenance of genome integrity. Multicellular organisms display tissue-specific, developmental and stress response strategies for telomerase suppression [59, 60]. In human somatic cells the inactivation of telomerase enzyme and maintenance of telomere length have been proposed to play a role as a tumor suppressor mechanism [61, 62]. This may also be needed for cell latency, differentiation, and death of some cell types [63]. However, collective telomere erosion limits the self-renewal ability of highly proliferative human cell lineages in the skin and blood [29]. The expression of TERC is universal while TERT expression is highly regulated in some organisms, especially in mammals. Many strategies have been proposed to control telomerase activity, because the enzyme can be regulated at various levels including expression level. For instance, the epigenetic modification of histones can modulate chromatin structure and the accessibility of the transcriptional machinery to regulatory regions of target genes. In this regard, numerous transcription factors, such as c-MYC, SP1, MAD1, and HIF-2a, have been shown to recruit either histone acetyltransferases or histone deacetylases to the TERT promoter to control TERT expression [64, 65]. However, the transcription expression is not constantly linked with the enzyme activity, which might result in transcription modulation failure [66]. Consequently, telomerase is expressed in embryonic stem cells, but TERT expression and telomerase activity are frequently very low or undetectable in somatic cells [67]. In contrast, telomerase activity seems to be high in most (85–90%) cancer cells [10, 46, 68]. Nevertheless, some cells that lack telomerase activity still exhibit a high level of hTERT transcription. In these cases, regulation at the level of alternative splicing leads to the skipping of exons that encode reverse transcriptase function [69]. In mice, the deletion of either TERC or TERT can result in telomere shortening, genomic instability, aneuploidy, telomeric fusion, and aging-related phenotypes [41, 70]. Therefore, telomerase dysfunction may lead to defects in various highly proliferative cells/tissues, ultimately leading to aging-related degenerative diseases [71]. The overexpression of TERT can dramatically increase the life span of mice in the background of the overexpression of tumor suppressor genes, such as p53, p16, and p19, indicating that TERT must have an anti-aging activity in mammals [4, 15, 72].

4.1 Telomere replication in the absence of telomerase

There exist some alternative mechanisms which are activated to maintain telomere length in the absence of telomerase activity. These mechanisms are principally based on recombination events that come into play to amplify or reorganize previously existing telomeric sequences [73, 74], and the mechanisms seem to be complementary to both the telomerase method and the method occupied in "retro transposition" [11]. The alternative mechanisms were first observed in budding yeasts that were able to survive and achieve telomere elongation despite lack of a functional telomerase [12]. Thereafter it was verified that this phenomenon is dependent on RAD52 (a protein involved in homologous recombination) [74]. Telomere lengths are also maintained by telomerase in most cancer cells, [46]. Reports have shown that approximately 10–15% of cancer cells elongate their telomeres by using one or more alternative mechanisms referred to as alternative telomere lengthening (ALT) [68, 75]. In the same way, immortalized cells can also elongate their telomeres using either telomerase [76] or ALT [77].

Other telomere-lengthening mechanisms also exist in the absence of telomerase activity. These mechanisms have been reviewed in details in previous reports [11]. The mosquito fly *Anopheles gambiae*, the vinegar fly *Drosophila melanogaster*, and some species of plants are other examples of organisms that use alternative telomere elongation mechanism by using recombination [11]. For instance, *Drosophila*, lacks telomerase activity and exhibits long tandem arrays composed of three non-LTR retrotransposons, HeT-A, TART, and TAHRE, instead of simple telomeric repeats unlike in most organisms. These were the first transposable elements revealed to play an important role in cell structure [11, 40, 78]. In *Trypanosoma brucei* (a haemoparasite), critically short telomeres generated by knocking out the TERT gene were stabilized by an unknown mechanism [79]. These short telomeres lack active transcriptional factors and tend to shorten more and more without leading to cell senescence due to their stability regardless of the absence of active telomerase enzyme [80, 81]. The mechanism by which these short telomeres are stabilized has not yet been revealed, but it is known that the telomerase-deficient strains switch variant surface genes (VSG) by duplicative gene conversion, which occurs more frequently than in wild-type strains and exhibit longer telomeres. Furthermore, it was observed that shorter chromosomes at no time underwent fusion and that telomere stabilization was sufficient to preserve genomic integrity, with no apparent effects on long-term population growth [82].

4.2 Methods of measuring telomerase activity

Methods used for the detection of telomerase activity can be divided into two major groups as described by Skvortsov et al. [83]: those based on direct detection of telomerase products, (**Table 1**) and those based on different systems of amplification of the signals from DNA yield from telomerase (**Table 2**). The methods discussed in this chapter (**Figures 3** and **4**) are suitable for testing telomerase activity in different types of samples such as; in protozoa, mammalian cells, mixed cellular populations, and tissues [83].

4.2.1 Methods containing the amplification of telomerase-synthesized DNA with modifications to the original TRAP

Telomeric repeat amplification protocols (TRAPs) are the most common methods employed for detection of telomerase activity which permit one to carry out semi-quantitative and quantitative analyses, by introducing some modifications [83].

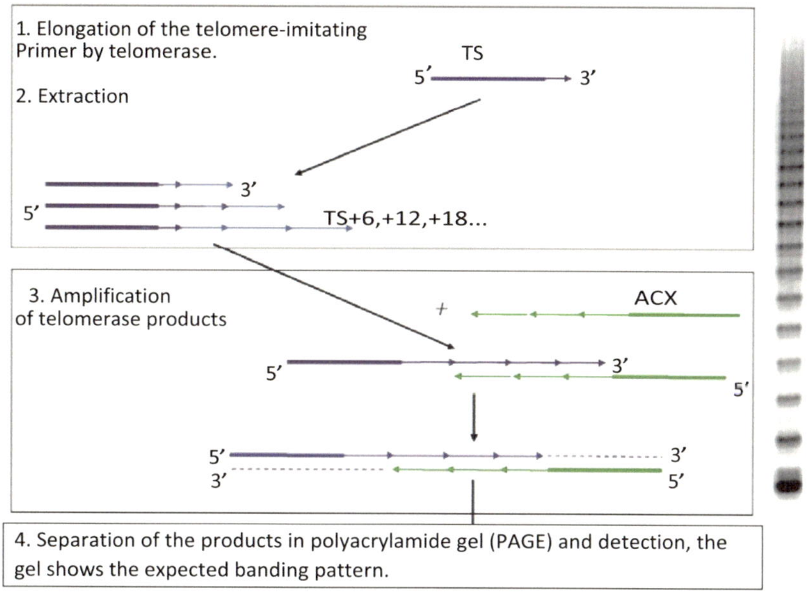

Figure 3.
Illustration showing the original telomeric repeat amplification protocol (TRAP) assay [83].

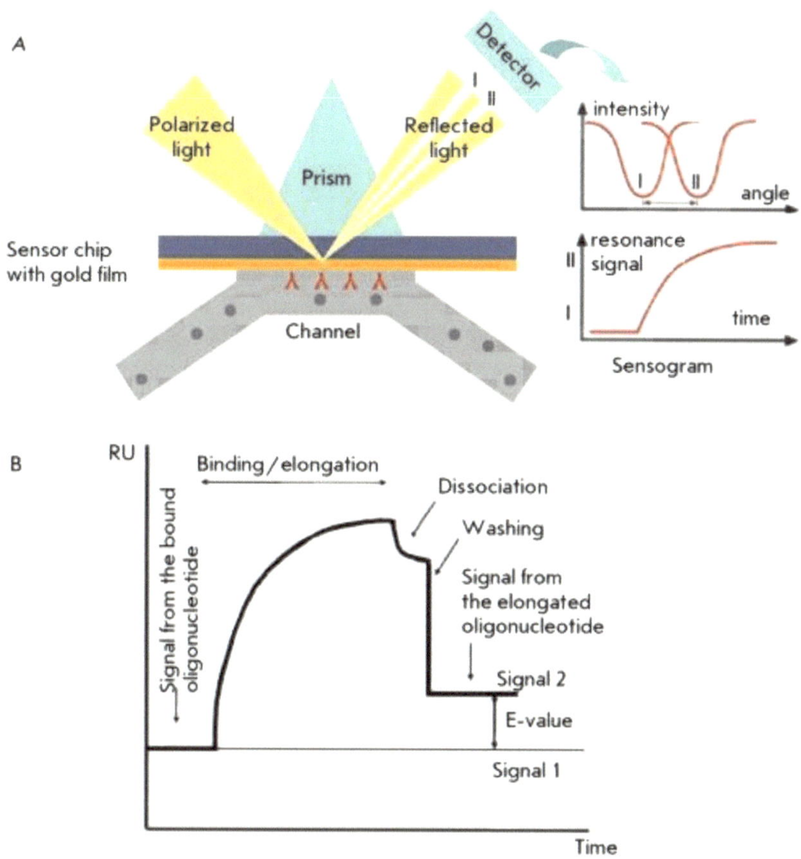

Figure 4.
Showing direct detection of telomerase by surface plasmon resonance (SPR) for detecting macromolecules; (A) sensogram corresponding to the general scheme and (B) SPR sensogram for telomerase activity detection. RU, resonance units. The difference between signals 1 and 2 represents DNA which was synthesized by telomerase [83].

Assay name	Description of modification of original assay	Benefits of modification	Potential limitation	References
Purification of telomerase-synthesized DNA and TRAP efficiency	This method involves the extraction of telomeric-synthesized DNA using modified magnetic beads where PCR inhibitors are eliminated or completely diluted. There are three main stages involved in the extraction process; the elongation of the substrate-imitating oligonucleotide by telomerase, the extraction of telomerase synthesized DNA using modified magnetic beads and finally the amplification step	Sensitivity towards PCR inhibitors is lower as compared with the standard TRAP method and its efficiency is slightly higher when analyzing tissue and other complex specimens	False-positive or false negative results can be obtained, and they may affect the validity of the diagnosis and prognosis of a disease. The nature of specimens, such as large volumes of fluid (blood, etc.) or presence of numerous normal cells, may also complicate the detection of telomerase activity	[83, 92]
Internal standards for TRAP	Here, the internal standards are amplified with the same primers as the telomerase-synthesized DNA. This permits the presence of Taq polymerase inhibitors (such as gem-containing compounds) in the samples to be analyzed and for performing of total control of PCR. The two most common standards used are of 36 and 150 bp in length	The 150 bp standard is more sensitive towards the Taq polymerase inhibitors present in the reaction mixture	There are two most common standards used, with the length of 36 and 150 bp. The 36 bp standard is excessively amplified if the specimens have a low telomerase activity, competes with the telomerase-synthesized DNA, and provides a false negative signal	[83, 90, 93]
"Two-primer" TRAP (TRAP) with an additional specific reverse primer	The telomerase-synthesized DNA is amplified by the use of two reverse primers with lengths of 20 (RP) and 38 (RPC3) nucleotides, in the presence of [3 H]TTP or [α-32P]dCTP	This modification of the standard TRAP is used to reduce false signals. Here there no electrophoretic analysis of PCR products but instead the total radioactivity is estimated as a decisive factor of telomerase activity		[83, 94, 95]
TRAP with fluorescence resonance energy transfer (FRET)	This method uses primers with an energy transfer (amplifluors) property instead of the normal primers used for amplification of the telomerase-synthesized DNA	The advantage of this modification is that the replacement of the telomerase substrate and reverse primer by amplifluors allows to achieve an increasingly high intensity of the fluorescent signal		[93, 96, 97]
TRAP with detection using a Scintillation proximity assay	This method uses scintillation proximity assay for the detection of amplified DNA instead of PAGE in TRAP. This modification helps to increase the detection rate of	In combination with the traditional TRAP, it helps increase the rate of detection of telomerase activity	The major disadvantage of this method is the use of tritium, and also like most methods for detecting TRAP products without PAGE, it is sensitive to PCR artifacts	[83, 98]

Assay name	Description of modification of original assay	Benefits of modification	Potential limitation	References
	telomerase activity. 5'-Biotin conjugated oligonucleotides are used as substrate in this method			
TRAP with detection using the hybridization protection assay (hybridization protection assay-TRAP)	In this method, after PCR, the amplified product is detected by the use of probes labeled with covalently bound acridine	This modification is safer and easier to detect telomerase product		[99]
TRAP with ELISA (enzyme-linked immunosorbent assay	In this method, after the amplification, concentration of DNA calorimetrically, enhancing the qualitative and semi-quantitative assessment of telomerase activity	TRAP-ELISA method is faster as compared with TRAP, which is based on the separation of the amplified DNA in gel. This makes it possible to use the TRAP-ELISA method in screening studies	One of the drawbacks of this method is the complexity in separating the telomerase-positive and telomerase-negative controls, which may result from the absence of internal controls and two steps of signal amplification	[100, 101]
TRAP with electrochemical detection	This method involves the treatment of the remaining PCR products after the initial PCR with 3 M HCl making it possible to determine dGMP (one of the products of the complete hydrolysis of the amplified DNA)	This modification has no special advantages over other methods in which no gel-electrophoresis or radioactive labeling are used	This method is more laborious and time consuming	[102]
TRAP with real-time PCR	This method employs the use of real time PCR which permits the simultaneous amplification of DNA and measurement of the amount of amplified products after each cycle. Combination of real time PCR with the standard TRAP makes it possible to obtain quantitative results and is suitable for analysis of large samples	This modification allows to obtain quantitative results. This modification is also suitable for studying telomerase inhibitors and analyzing large specimen series. The advantages of this method include; a high rate of detection of telomerase-synthesized DNA upon flow analysis and enhanced specificity due to the use of corresponding probes	Over-estimation of telomerase activity and leveling of small differences in the activity is possible without the assessment of the amount of real-time PCR products, due to saturation of the PCR reaction in the final stages. Also, there is the possibility of dimer formation resulting in false-positive signal	[83, 87, 103–105]
In situ TRAP	In this method, FITC-labeled primers (direct and reverse) are used. Here the fluorescence intensity and its localization in nucleus or cytoplasm are used to determine telomerase activity in separate cell types within a mixture. High telomerase activity in urogenital and bronchial lavages manifests	This modification allows for a semi-quantitative determination of telomerase activity and its localization in isolated cells of tissues and cell suspensions	This method is only appropriate for cancer cells and not suitable to study cell senescence	[106–111]

Assay name	Description of modification of original assay	Benefits of modification	Potential limitation	References
	itself through bright fluorescence of the nuclei of malignant cells. The signal from cytoplasm is detected in granulocytes and macrophages			
TRAP on microchips	This method involves the combination of two primer—TRAP and binding of PCR products on chips, followed by probe hybridization and detection using different fluorescent labels to determine telomerase synthesized DNA and the internal standard; for example, Cy3 for the telomerase-synthesized DNA being amplified and Cy5, for the standard	The modification is that; the introduction of microchips and hybridization using different probes for the detection of telomerase synthesized DNA		[83]
Transcription amplification of telomerase-synthesized DNA	In this modification PCR is replaced by transcription amplification, this is to increase the amount of telomerase-synthesized DNA. Combined with "hybridization protection," this method allows to determine the telomerase activity within short time period. The method's sensitivity allows to detect telomerase activity in specimens consisting of 1–1000 cells and also sensitive to the presence of RNases	The major advantage of this method over TRAP with PCR amplification is that no heating of a specimen is required upon amplification and that the specific Taq polymerase inhibitors are neglected. This method makes it possible to semi-quantitatively determine the telomerase activity in tissue and cell line extracts		[83, 87, 112]

Table 1.
Different modifications of TRAP method for telomerase activity detection.

DOI: http://dx.doi.org/10.5772/intechopen.89506

Assay name	Description of modification or original assay	Benefits of modification	Potential limitation or disadvantage	References
Direct incorporation of a radioactively labeled substrate	This method is based on the ability of telomerase to elongate oligonucleotides in the presence of dNTP. The oligonucleotide or the dNTPs are radioactive labeled allowing the detection of telomerase activity electrophoretically by its incorporation into the telomerase-synthesized DNA	The modification allows for qualitative determination of the activity and processivity of telomerase in cell line extracts. Absence of artifacts associated with PCR. The telomerase-synthesized DNA can be immediately observed in the gel. Its size and amount can be estimated	The major disadvantage of this method includes the use of large amounts of radioactive isotopes with longer exposure time and insufficient sensitivity	[83, 113]
Determination of telomerase-synthesized DNA by changes in surface plasmon resonance	Surface plasmon resonance (SPR) is used to detect telomerase activity as the corresponding elongation of a telomere-imitating oligonucleotide, using a biosensor (5'-biotin-conjugated) oligonucleotides, which contain telomeric repeats, these are in situ immobilized on the surface of a dextrane sensor pre-treated with streptavidin on a BIACORE instrument	This method allows to quantitatively determine telomerase activity in tissue and cell line extracts and collect data on the kinetics of the reaction, demonstrating the binding and dissociation of telomerase from the substrate	The necessity of the BIACORE system and biotin-conjugated primers	[83, 114, 115]
Oligo-modified magnetic particles and NMR	This method involves the use of magnetic particles of an iron oxide which are modified with oligonucleotides that are complementary to telomeric repeats. These particles bind to the telomerase-synthesized repeats due to complementary interactions and form extended linear structures (MRS complexes). The local distortion of the magnetic field increases on nanoparticles in ordered ensembles, whereas the nonordered nanoparticles provide a considerably lower magnetic effect	This modification made it possible to analyze large number of specimens within short period of time (in several 10 of minutes) with high sensitivity. There also absence of artifacts which were associated with PCR. Radioactive label and PAGE are not required	In addition to the common instruments and reagents, these analyses require a plate NMR-spectrometer and a specimen of oligonucleotide-modified nanoparticles which are expensive	[83, 116]
Quartz crystal microbalance technique	An Au-quartz resonator can be used for a microgravitometrical analysis of telomerase activity, according to the quartz crystal microbalance technique. Here, in this method, the telomere-imitating oligonucleotide is bound to the surface of the sensor and is elongated by telomerase	The method allows for the quantitative determination of the telomerase activity in tissue and cell line extracts. The sensitivity threshold is up to 3300-cell extracts of the telomerase-positive cell line at a high rate	A frequency analyzer and an Au-quartz crystal need to be provided in this method. Also, the identification of artifact signals is complicated	[117]

Assay name	Description of modification or original assay	Benefits of modification	Potential limitation or disadvantage	References
Biobarcode assay for telomerase activity detection	In the original biobarcode system, magnetic particles bind to a target, which in turn binds to nanoparticles covalently modified with an oligonucleotide due to antigen–antibody interactions. DNA nanoparticles consist of gold nanoparticles and oligonucleotides of two types; one of these can form a duplex with the telomerase-synthesized DNA	This modification makes it most sensitive for direct detection of telomerase activity without amplification of the telomerase-synthesized DNA. Whereas the second cannot. Because of this, the probability of binding another DNA-target to the same nanoparticle is reduced. The electroactive complex $[Ru(NH_3)_6]^{3+}$, which is capable of binding to negatively charged DNA chains due to electrostatic interactions, is used for detection	A modified electrode and nanoparticles coated with oligonucleotides of two types are required	[118]
Optical biosensors assay	The principle of this method is similar to that of SPR. The method is based on the fact that upon binding of a target, the refraction index on the sensor's surface changes in proportion to the amount of bound targets	In this modification a cassette consisting of three oligonucleotides helps to avoid steric impediments. Phosphate groups covalently interact with the surface via the 5′-end of an oligonucleotide. Then, an oligonucleotide containing a short noncomplementary region on its 3′-end complementarily binds to the immobilized DNA. The prominent 3′-end of the DNA is modified with phosphorothioate, which enhances the affinity of telomerase-primer binding by a factor of 10	Phosphorothioate-modified oligonucleotides and a special optosensor are required. The low sensitivity for clinical samples	[83, 119]
Quantum dots telomerase detection	This method is based on a telomere-imitating oligonucleotide modified at its end with the thio group is attached to a nanoparticle (quantum dot). This quantum dot is capable of fluorescing by absorbing a quantum with a wavelength of λ_1 (400 nm) and emitting a quantum with a wavelength of λ_1' (560 nm)	This modification allows for a quantitative assessment of the telomerase activity	The disadvantage is that the sensitivity threshold is approximately 10,000 HeLa cells which is not enough for an analysis of clinical materials	[120]
An on-chip nanowire sensor assay	This method involves the use a sensor chip (a transistor comprising antibody-coated silicon nanowires with aldehydes groups on their surface, to which monoclonal antibodies can be linked) to determine telomerase activity, the transistor nanowire is modified by telomere-imitating oligonucleotides instead of antibodies.	This modification allows to determine telomerase activity in cell line extracts even though, it has not been tested on clinical materials. Its sensitivity is sixty five (65) T293 cells and the major advantage of this method is that a large number of different analyses of one specimen can be carried out on a single chip,	Here a transistor chip and the equipment to analyze it are required	[121, 122]

Assay name	Description of modification or original assay	Benefits of modification	Potential limitation or disadvantage	References
	After the introduction of telomerase and dNTP, the oligonucleotides bound on the surface elongate, which results in change in the conductivity of the transistor to which the oligonucleotides are bound	which can be a set of sensors with respect to various markers. Moreover, the stages of telomerase binding and dissociation can be observed		
Bioluminescence method	Bioluminescence method is used for the determination of telomerase activity based on the fact that the telomerase-catalyzed elongation of the telomere-imitating oligonucleotide is accompanied by the cleavage of pyrophosphate; its amount is determined luminometrically	The advantage of the method is the linear dependence of the signal on the amount of telomerase-synthesized DNA, combined with high efficiency. The sensitivity and specificity of this method can be comparable with the sensitivity of TRAP-ELISA	A luciferase system for bioluminescence detection and a luminometer are required	[123]
Electrochemiluminescence method	Telomerase activity can also be determined by electrochemiluminescence (luminescence upon electrolysis). In this method, the 5′-biotin-conjugated primer is elongated by telomerase followed by incubation with a suspension of magnetic beads modified with avidin	This method allows for the quantitative determination of telomerase activity in samples that contain at least 500 HeLa cells. It provides an appreciably high signal/noise ratio due to the stage of magnetic bead extraction; however, it has not been tested on clinical materials	Difficult to synthesize a sample, requirements to the equipment	[83, 124]
FRET and total internal reflection fluorescence microscopy method	This is a FRET-based method that is intended to distinguish the single-letter synthesis (a nonprocessive method of synthesis) and the beginning of synthesis of the second DNA repeat (a conditionally processive method of synthesis) by individual complexes of *Tetrahymena thermophila* telomerase. Biotin-conjugated primers (TG)8 T2 G4 T2 were used on streptavidin-coated quartz slides which were treated with telomerase-containing extract in the presence of dGTP and ddTTP	The method allows to identify individual signals from the elongation of the primer by telomerase and can be combined with FRET-based methods to investigate telomerase structure	This method can be used to determine only the first 1.5 telomerase-synthesized repeats; that is, it does not reveal total telomerase activity. The method has been tested only on *T. thermophila* telomerase and has been aimed only at solving research problems till date	[83, 125]

Table 2.
Methods for direct detection of telomerase-synthesized DNA.

Assay name	Description of method	Advantage of assay	Potential limitation/disadvantage	References
Telomere restriction fragment (TRF) analysis	Telomere restriction fragments (TRF) involves Southern blot hybridization using probes against telomere repeats used for the analysis	This technique is widely used and it requires no special reagents or equipment	Quantification is very difficult. It requires large number of cells ($\sim10^6$). This method provides an estimate of the average telomere length per sample and produces subtelomeric polymorphism	[52, 128]
qPCR	The qPCR technique was first introduced by RM Cawthon in 2002. The method involve detection of telomeric DNA with fluorescent signals (T) using partially mismatched primers in a 96-well format on real time-quantitative PCR (qPCR) platform. The measurement of the telomeric DNA is normalized with a single-copy housekeeping gene (S) that is amplified in same sample in a different plate and T/S ratio (measure of relative TL) is computed	In qPCR technique, relative TL can be assessed within a short time and small amount of DNA (20 ng per reaction) is required. This method is widely used in epidemiological studies that involve large number of samples	Studies have shown that there is wide range of CVs (2–28%) for measurement of TL by qPCR which suggests that repeatability is a concern with the qPCR technique. Proper optimization of qPCR conditions is required to reduce variability	[129, 130, 133–135]
Monochrome multiplex qPCR (MMqPCR)	It was established by the same person as qPCR. This is an improved version of the qPCR method in which both telomeric DNA and single-copy gene are amplified in same well of a plate	In this method there is less variability compared with monoplex qPCR and has lesser sample requirement	Just like the qPCR proper optimization of experimental conditions is required	[130]
Flow-fluorescence *in situ* hybridization (FISH) assay	Telomere length analysis by FISH is based on the specific labeling of telomeres with fluorescent peptide nucleic acid (PNA) oligonucleotide probes	Increased sensitivity and specificity. It has ability to measure telomere length at the single-cell level	This method requires expensive equipment which may not be found in many laboratories and it is not possible to assess TL in tissues (histological samples) and stored samples. The assay typically requires cells to be at replicating stage to analyze chromosomes in metaphases rather than interphase cells	[131]
Single telomere length analysis (STELA)	It is a ligation PCR-based method	No specialized equipment is required and it requires very limited starting material	The single telomere length analysis method is also labor intensive and is again not appropriate for the analysis of large number of samples	[132]

Table 3.
Different methods for measurement of telomere length.

These modifications may include: rising of the rate of analysis, substitution of the radioactive label by nonlabeled compounds, the decrease in the amount of side products, and so on. A number of modifications still make it possible to detect telomerase activity within a single cell [84]. The TRAP consists of three main stages: primer elongation, amplification of telomerase-synthesized DNA, and finally, the detection step. In the elongation step, telomeric repeats are added to the telomere-imitating oligonucleotide (TS) by telomerase found in the cell extract. Next, PCR-amplification of telomerase synthesized DNA is carried out using definite primers (telomere-imitating and reverse primers). At this stage, different labels such as radioactive or fluorescent labels can be incorporated into the synthesized DNA. This step is then followed by detection through separation of PCR products by gel electrophoresis and imaging) [83]. The original TRAP assay has several drawbacks. Initially, the CX oligonucleotide, which complementarily overlaps with TS for several base pairs (bp), is used in the amplification of PCR products which results in the primer dimer formation as a result of the interaction between primers and products [83]. The use of optimal ACX primer with the noncomplementary TS end can lead to appearance of background signal during the analysis of concentrated tissue extracts from tumor [85]. The use of an oligonucleotide TSG4 which can also be added to the TRAP mixture in order to evaluate the effect of duplex-stabilizing inhibitors. This oligonucleotide does not require the synthesis of several repeats by telomerase before the inhibitor begins its action [86]. Various nucleotides used in TRAP assay were discussed in more detail [87]. In addition, when PCR is used for signal amplification, the PCR inhibitors contained in the specimen can alter the results of telomerase activity detection [83]. Previously, in the TRAP method, PCR products were detected in polyacrylamide gel (PAGE) with respect to the radioactive label used, it can be introduced using a radioactively labeled primer or incorporated into the DNA during the preparation of the PCR reaction. This method allows performing a qualitative assessment of the activity and processivity of telomerase in cells and tissue extracts [83]. In the second stage of TRAP, the PCR product allows to obtain an amount of DNA sufficient for gel staining, for instance, use of ethidium bromide [88], silver nitrite [89] and SYBR Green [90] and its analogs (which has sensitivity equal to that of radioactive label [91], while mutagenicity is considerably higher than when ethidium bromide is used) (**Tables 1** and **2**).

4.3 Methods used for measurement of telomere length (TL)

Several researchers have shown interest in measuring telomere length (TL) accurately and efficiently so as to understand both the fundamental biology of telomere maintenance as well as factors which contributes significantly to accelerated TL attrition. Tarik et al. [126, 127] have described different techniques which were developed for telomere length measurement (**Table 3**).

5. Conclusions

Telomerase is a ribonucleoprotein complex, composed of a reverse transcriptase enzyme catalytic subunit and a long non-coding RNA that contains the template sequence for telomere synthesis and is required for linear chromosome maintenance in most eukaryotes. Telomerase is a high-molecular weight RNP complex that consists of two major components: TERC and a TERT. It was found that in humans; only the TERT and TERC components of the telomerase are required for its activity *in vitro*, even though some proteins which have regulatory function are also essential for the catalytic function of telomerase *in vivo*. It was estimated that about 32 different proteins are involved with human telomerase *in vivo* which maintain its functionality and only some of these proteins are phylogenetically preserved. The molecular chaperones Hsp90 and p23, which directly associate with hTERT, are present in rabbit reticulocyte lysates and are necessary for telomerase activity. Studies suggest that in *Tetrahymena* the proteins p80 and p95 are not core telomerase components and may be separate ribonucleoproteins that copurified non-specifically with telomerase. Telomerase remains active during early embryonic development but expression declines after the blastocyst stage and can no longer be detected in neonatal somatic cells.

Until recently, the activity of telomerase was thought to be controlled by limiting access to the telomeres but it is regulated by some protein complexes such as the shelterin complexes. Nevertheless, the cumulative information given in this chapter show that the events involved in telomerase recruitment and its activation are separate. Although, the organizational biology and biochemistry responsible for the process of telomerase activation is still unknown which could be an important focus area in future research. Telomerase activity is highly expressed in embryonic germ cells, testes, ovaries and in some cancer cells but its activity is low or absent in somatic cells. The expression of telomerase activity in cells indicates replicative capability of that cell and this involves several factors which regulate the telomerase activity. Towards understanding the biology of telomere, several methods have been designed to measure the telomerase activity and the telomere length. TRAP was the initial method deployed for the measurement of telomerase where the amplified product is detected using gel electrophoresis. There are several other modifications to the original TRAP which has more advantages, such as the qPCR amplification method which uses less concentration of primers and permits quantitative determination of synthesized DNA. However, proper optimization of qPCR conditions is required to achieve reproducibility of this method. There are various methods which have been established for measurement of telomere length (TL) and these includes: (i) terminal restriction fragment (TRF) analysis (the gold standard), (ii) flow-FISH cytometry of cells following hybridization with fluorescent peptide nucleic acid (PNA) probes, (iii) quantitative fluorescence in situ hybridization (FISH) with fluorescent telomere PNA probes and (iv) qPCR assay. Monochrome multiplex qPCR (MMqPCR) was also established which is an improved version of the qPCR method in which both telomeric DNA and single-copy gene are amplified in a same well of a plate which require lesser sample and shows less variability. Studies have shown that there is wide range of CVs (2–28%) for measurement of TL by qPCR which suggests that repeatability is a concern with the qPCR technique. Therefore, proper optimization of qPCR protocols is required to reduce variability in the results.

Apart from the TRF assay all other methods have the problem of generating a relative measure of TL. While the qPCR technique has more advantage where it requires small amounts of DNA, less time consuming and can easily be performed in high-throughput format which makes it possible to analyze large epidemiological samples.

Conflict of interest

The authors have declared that there is no conflict of interest.

Author details

Prem Singh Yadav[1*] and Abubakar Muhammad Wakil[1,2,3]

1 Division of Animal Physiology and Reproduction, ICAR, Central Institute for Research on Buffaloes (CIRB), Hisar, Haryana, India

2 Department of Veterinary Physiology and Biochemistry, University of Maiduguri, Maiduguri, Nigeria

3 Department of Veterinary Physiology and Biochemistry, College of Veterinary Sciences, Lala Lajpat Rai University of Veterinary and Animal Sciences, Hisar, Haryana, India

*Address all correspondence to: prem.yadav@icar.gov.in

References

[1] Armstrong CA, Tomita K. Fundamental mechanisms of telomerase action in yeasts and mammals: Understanding telomeres and telomerase in cancer cells. Open Biology. 2017;**7**:160338

[2] Greider CW, Blackburn EH. Identification of a specific telomere terminal transferase activity in *Tetrahymena* extracts. Cell. 1985;**43**: 405-413

[3] Burgstaller JP, Brem G. Aging of cloned animals: A mini-review. Gerontology. 2017;**63**:417-425

[4] Tian X, Doerig K, Park R, Can Ran Qin A, Hwang C, Neary A, et al. Evolution of telomere maintenance and tumour suppressor mechanisms across mammals. Philosophical Transactions of the Royal Society B. 2018;**373**:20160443

[5] Harrington L. Making the most of a little: Dosage effects in eukaryotic telomere length maintenance. Chromosome Research. 2005;**13**: 493-504

[6] Hiyama E, Hiyama K. Telomere and telomerase in stem cells. British Journal of Cancer. 2007;**96**:1020-1024

[7] Liu Y, Kha H, Ungrin M, Robinson MO, Harrington L. Preferential maintenance of critically short telomeres in mammalian cells heterozygous for mTert. Proceedings of the National Academy of Sciences of the United States of America. 2002;**99**: 3597-3602

[8] Autexier C, Lue NF. The structure and function of telomerase reverse transcriptase. Annual Review of Biochemistry. 2006;**75**:493-517

[9] Greider CW. Telomere length regulation. Annual Review of Biochemistry. 1996;**65**:337-365

[10] Kim NW, Piatyszek MA, Prowse KR, Harley CB, West MD, Ho PL, et al. Specific association of human telomerase activity with immortal cells and cancer. Science. 1994;**266**(5193):2011-2015

[11] Pardue ML, DeBaryshe PG. Retrotransposons provide an evolutionarily robust nontelomerase mechanism to maintain telomeres. Annual Review of Genetics. 2003;**37**: 485-511

[12] Giardini MA, Segatto M, da Silva MS, Nunes VS, Cano MIN. Telomere and telomerase biology. Progress in Molecular Biology and Translational Science. 2014;**125**:1-40

[13] Chen Q, Ijpma A, Greider CW. Two survivor pathways that allow growth in the absence of telomerase are generated by distinct telomere recombination events. Molecular and Cellular Biology. 2001;**21**:1819-1827

[14] Lundblad V, Blackburn EH. An alternative pathway for yeast telomere maintenance rescues est1-senescence. Cell. 1993;**73**:347-360

[15] Tomas-Loba A, Flores I, Fernandez-Marcos PJ, et al. Telomerase reverse transcriptase delays aging in cancer-resistant mice. Cell. 2008;**135**:609-622

[16] Blackburn EH. Telomeres and their synthesis. Science. 1990;**249**:489-490

[17] Wu RA, Upton HE, Vogan JM, Collins K. Telomerase mechanism of telomere synthesis. Annual Review of Biochemistry. 2017;**86**:439-460

[18] Min J, Wright WE, Jerry WS. Alternative lengthening of telomeres mediated by mitotic DNA synthesis engages break-induced replication processes. Molecular and Cellular Biology. 2017;**37**(20):e00226-e00217

[19] Sandin S, Rhodes D. Telomerase structure. Current Opinion in Structural Biology. 2014;**25**:104-110

[20] Chen JL, Greider CW. Functional analysis of the pseudoknot structure in human telomerase RNA. Proceedings of the National Academy of Sciences of the United States of America. 2005;**102**(23):8080-8085; discussion 8077-9. DOI: 10.1073/pnas.0502259102

[21] Chen JL, Greider CW. An emerging consensus for telomerase RNA structure. Proceedings of the National Academy of Sciences of the United States of America. 2004;**101**(41): 14683-14684. DOI: 10.1073/pnas.0406204101. Erratum in: Proceedings of the National Academy of Sciences of the United States of America. 2004;101(46):16391

[22] Chen JL, Greider CW. Telomerase RNA structure and function: Implications for dyskeratosis congenita. Trends in Biochemical Sciences. 2004; **29**(4):183-192

[23] Chen JL, Greider CW. Template boundary definition in mammalian telomerase. Genes & Development. 2003;**17**(22):2747-2752

[24] Chen JL, Greider CW. Determinants in mammalian telomerase RNA that mediate enzyme processivity and cross-species incompatibility. The EMBO Journal. 2003;**22**(2):304-314

[25] Chen JL, Opperman KK, Greider CW. A critical stem-loop structure in the CR4-CR5 domain of mammalian telomerase RNA. Nucleic Acids Research. 2002;**30**(2): 592-597

[26] Chen JL, Blasco MA, Greider CW. Secondary structure of vertebrate telomerase RNA. Cell. 2000;**100**(5): 503-514

[27] Harrington L. Biochemical aspects of telomerase function. Cancer Letters. 2003;**194**:139-154

[28] Cohen SB, Graham ME, Lovrecz GO, Bache N, Robinson PJ, Reddel RR. Protein composition of catalytically active human telomerase from immortal cells. Science. 2007;**315**: 1850-1853

[29] Collins K. The biogenesis and regulation of telomerase holoenzymes. Nature Reviews. Molecular Cell Biology. 2006;**7**:484-494

[30] Lingner J, Cech TR, Hughes TR, Lundblad V. Three ever shorter telomere (EST) genes are dispensable for in vitro yeast telomerase activity. Proceedings of the National Academy of Sciences of the United States of America. 1997;**94**:11190-11195

[31] Wong JMY, Collins K. Telomere maintenance and disease. Lancet. 2003; **362**:983-988

[32] Fu D, Collins K. Purification of human telomerase complexes identifies factors involved in telomerase biogenesis and telomere length regulation. Molecular Cell. 2007;**28**: 773-785

[33] Smogorzewska A, de Lange T. Regulation of telomerase by telomeric proteins. Annual Review of Biochemistry. 2004;**73**:177-208

[34] Nandakumar J, Cech TR. Finding the end: Recruitment of telomerase to telomeres. Nature Reviews. Molecular Cell Biology. 2013;**14**:69-82

[35] Gilson E, Geli V. How telomeres are replicated. Nature Reviews. Molecular Cell Biology. 2007;**8**:825-838

[36] Londono-Vallejo JA, Wellinger RJ. Telomeres and telomerase dance to the rhythm of the cell cycle. Trends in Biochemical Sciences. 2012;**37**:391-399

[37] Stewart JA, Wang F, Chaiken MF, et al. Human CST promotes telomere duplex replication and general replication restart after fork stalling. The EMBO Journal. 2012;**31**: 3537-3549

[38] Watson JD. Origin of concatemeric T7 DNA. Nature: New Biology. 1972; **239**:197-201

[39] Cong YS, Wright WE, Shay JW. Human telomerase and its regulation. Microbiology and Molecular Biology Reviews: MMBR. 2002;**66**(3): 407-425

[40] Mason JM, Frydrychova RC, Biessmann H. Drosophila telomeres: An exception providing new insights. BioEssays. 2008;**30**:25-37

[41] Liu Y, Snow BE, Hande MP, et al. The telomerase reverse transcriptase is limiting and necessary for telomerase function in vivo. Current Biology. 2000; **10**:1459-1462

[42] Szostak JW, Blackburn EH. Cloning yeast telomeres on linear plasmid vectors. Cell. 1982;**29**:245-255

[43] Zakian VA. Telomeres: Beginning to understand the end. Science. 1995;**270**: 1601-1607

[44] Hiyama K, Hirai Y, Kyoizumi S, Akiyama M, Hiyama E, Piatyszek MA, et al. Activation of telomerase in human lymphocytes and hematopoietic progenitor cells. Journal of Immunology. 1995;**155**:3711-3715

[45] Jafri MA, Ansari SA, Alqahtani MH, Shay JW. Roles of telomeres and telomerase in cancer, and advances in telomerase-targeted therapies. Genome Medicine. 2016;**8**:69

[46] Shay JW, Bacchetti S. A survey of telomerase activity in human cancer. European Journal of Cancer. 1997;**33**: 787-791

[47] Bodnar AG et al. Extension of life-span by introduction of telomerase into normal human cells. Science. 1998;**279**: 349-352

[48] de Lange T, Shiue L, Myers RM, Cox DR, Naylor SL, Killery AM, et al. Structure and variability of human chromosome ends. Molecular and Cellular Biology. 1990;**10**:518-527

[49] Hastie ND, Dempster M, Dunlop MG, Thompson AM, Green DK, Allshire RC. Telomere reduction in human colorectal carcinoma and with ageing. Nature. 1990;**346**:866-868

[50] Xu L, Blackburn EH. Human cancer cells harbor T-stumps, a distinct class of extremely short telomeres. Molecular Cell. 2007;**28**:315-327

[51] Teixeira MT, Arneric M, Sperisen P, Lingner J. Telomere length homeostasis is achieved via a switch between telomerase-extendible and nonexpendable states. Cell. 2004;**117**: 323-335

[52] Harley CB, Futcher AB, Greider CW. Telomeres shorten during ageing of human fibroblasts. Nature. 1990;**345**(6274):458-460

[53] Wright WE, Piatyszek MA, Rainey WE, Byrd W, Shay JW. Telomerase activity in human germline and embryonic tissues and cells. Developmental Genetics. 1996;**18**: 173-179

[54] Gilchrist GC, Kurjanowicz P, Mereilles FV, King WA, LaMarre J. Telomere length and telomerase activity in bovine pre-implantation embryos in vitro. Reproduction in Domestic Animals. 2015;**50**:58-67. DOI: 10.1111/rda.12449

[55] Turner S, Wong HP, Rai J, Hartshorne GM. Telomere lengths in human oocytes, cleavage stage embryos

and blastocysts. Molecular Human Reproduction. 2010;**16**:685-694

[56] Ramirez RD, Wright WE, Shay JW, Taylor RS. Telomerase activity concentrates in the mitotically active segments of human hair follicles. The Journal of Investigative Dermatology. 1997;**108**:113-117

[57] Greenwood MJ, Lansdorp PM. Telomeres, telomerase, and hematopoietic stem cell biology. Archives of Medical Research. 2003;**34**: 489-495

[58] Hayflick L, Moorhead PS. The serial cultivation of human diploid cell strains. Experimental Cell Research. 1961;**25**: 585-621

[59] Djojosubroto MW, Choi YS, Lee HW, Rudolph KL. Telomeres and telomerase in aging, regeneration and cancer. Molecules and Cells. 2003;**15**: 164-175

[60] Schaetzlein S, Rudolph KL. Telomere length regulation during cloning, embryogenesis and ageing. Reproduction, Fertility, and Development. 2005;**17**:85-96

[61] Sharpless NE, DePinho RA. Telomeres, stem cells, senescence, and cancer. The Journal of Clinical Investigation. 2004;**113**:160-168

[62] Shay JW, Wright WE. Senescence and immortalization: Role of telomeres and telomerase. Carcinogenesis. 2005; **26**:867-874

[63] Blackburn EH. Cell biology: Shaggy mouse tales. Nature. 2005;**436**:922-923

[64] Hou M, Wang X, Popov N, et al. The histone deacetylase inhibitor trichostatin A derepresses the telomerase reverse transcriptase (hTERT) gene in human cells. Experimental Cell Research. 2002;**274**: 25-34

[65] Xu D, Popov N, Hou M, et al. Switch from Myc/Max to Mad1/Max binding and decrease in histone acetylation at the telomerase reverse transcriptase promoter during differentiation of HL60 cells. Proceedings of the National Academy of Sciences of the United States of America. 2001;**98**:3826-3831

[66] Gladych M, Wojtyla A, Rubis B. Human telomerase expression regulation. Biochemistry and Cell Biology. 2011;**89**:359-376

[67] Blasco MA. Telomeres and human disease: Ageing, cancer and beyond. Nature Reviews. Genetics. 2005;**6**: 611-622

[68] Cesare AJ, Reddel RR. Alternative lengthening of telomeres: Models, mechanisms and implications. Nature Reviews. Genetics. 2010;**11**:319-330

[69] Ulaner GA, Hu JF, Vu TH, Giudice LC, Hoffman AR. Telomerase activity in human development is regulated by human telomerase reverse transcriptase (hTERT) transcription and by alternate splicing of hTERT transcripts. Cancer Research. 1998;**58**: 4168-4172

[70] Blasco MA, Rizen M, Greider CW, Hanahan D. Differential regulation of telomerase activity and telomerase RNA during multi-stage tumorigenesis. Nature Genetics. 1996;**12**:200-204

[71] Lu W, Zhang Y, Liu D, Songyang Z, Wan M. Telomeres-structure, function, and regulation. Experimental Cell Research. 2013;**319**:133-141

[72] Rahman R, Latonen L, Wiman KG. hTERT antagonizes p53-induced apoptosis independently of telomerase activity. Oncogene. 2005;**24**:1320-1327

[73] Henson JD, Neumann AA, Yeager TR, Reddel RR. Alternative lengthening of telomeres in mammalian cells. Oncogene. 2002;**21**:598-610

[74] Teng SC, Zakian VA. Telomere-telomere recombination is an efficient bypass pathway for telomere maintenance in *Saccharomyces cerevisiae*. Molecular and Cellular Biology. 1999;**19**: 8083-8093

[75] Reddel RR. Alternative lengthening of telomeres, telomerase, and cancer. Cancer Letters. 2003;**194**:155-162

[76] Counter CM, Avilion AA, LeFeuvre CE, et al. Telomere shortening associated with chromosome instability is arrested in immortal cells which express telomerase activity. The EMBO Journal. 1992;**11**:1921-1929

[77] Bryan TM, Englezou A, Gupta J, Bacchetti S, Reddel RR. Telomere elongation in immortal human cells without detectable telomerase activity. The EMBO Journal. 1995;**14**:4240-4248

[78] Pardue ML, DeBaryshe PG. Drosophila telomeres: A variation on the telomerase theme. Fly (Austin). 2008;**2**: 101-110

[79] Dreesen O, Cross GA. Telomerase-independent stabilization of short telomeres in *Trypanosoma brucei*. Molecular and Cellular Biology. 2006; **26**:4911-4919

[80] Dreesen O, Cross GA. Consequences of telomere shortening at an active VSG expression site in telomerase-deficient *Trypanosoma brucei*. Eukaryotic Cell. 2006;**5**: 2114-2119

[81] Dreesen O, Li B, Cross GA. Telomere structure and function in trypanosomes: A proposal. Nature Reviews. Microbiology. 2007;**5**:70-75

[82] Hovel-Miner GA, Boothroyd CE, Mugnier M, Dreesen O, Cross GA, Papavasiliou FN. Telomere length affects the frequency and mechanism of antigenic variation in *Trypanosoma brucei*. PLoS Pathogens. 2012;**8**: e1002900

[83] Skvortsov DA, Zvereva ME, Shpanchenko OV, Dontsova OA. Assays for detection of telomerase activity. Acta Naturae. 2011;**3**(1):48-68

[84] Wright WE, Shay JW, Piatyszek MA. Modifications of a telomeric repeat amplification protocol (TRAP) result in increased reliability, linearity and sensitivity. Nucleic Acids Research. 1995;**23**(18): 3794-3795

[85] Skvortsov DA, Gasparyan NM, Rubtsova MP, et al. Telomerase as a potential marker for early diagnosing cervical carcinoma. Dokl Biochem Biophys. 2006;**408**:158-160

[86] Gomez D, Mergny JL, Riou JF. Detection of telomerase inhibitors based on G-quadruplex ligands by a modified telomeric repeat amplification protocol assay. Cancer Research. 2002;**62**(12): 3365-3368

[87] Saldanha SN, Andrews LG, Tollefsbol TO. Analysis of telomerase activity and detection of its catalytic subunit, hTERT. Analytical Biochemistry. 2003;**315**(1):1-21

[88] Gan Y, Lu J, Johnson A, Wientjes MG, Schuller DE, Au JL. A quantitative assay of telomerase activity. Pharmaceutical Research. 2001; **18**(4):488-493

[89] Dalla Torre CA, Maciel RM, Pinheiro NA, Andrade JA, De Toledo SR, Villa LL, et al. TRAP-silver staining, a highly sensitive assay for measuring telomerase activity in tumor tissue and cell lines. Brazilian Journal of Medical and Biological Research. 2002; **35**(1):65-68

[90] Holt SE, Norton JC, Wright WE, Shay JW. Comparison of the telomeric repeat amplification protocol (TRAP) to

the new TRAP-eze telomerase detection kit. Methods in Cell Science. 1996;**18**: 237-248

[91] Skvortsov DA, Zvereva ME, Pavlova LS, Petrenko AA, Kisseljov FL, Dontsova OA. Vestn. MSU. 2010;**65**(3): 165-169

[92] Gollahon LS, Holt SE. Alternative methods of extracting telomerase activity from human tumor samples. Cancer Letters. 2000;**159**(2):141-149

[93] Uehara H, Nardone G, Nazarenko I, Hohman RJ. Detection of telomerase activity utilizing energy transfer primers: Comparison with gel- and ELISA-based detection. BioTechniques. 1999;**26**(3):552-558

[94] Szatmari I, Tokes S, Dunn CB, Bardos TJ, Aradi J. Modified telomeric repeat amplification protocol: A quantitative radioactive assay for telomerase without using electrophoresis. Analytical Biochemistry. 2000;**282**(1):80-88

[95] Szatmari I, Aradi J. Telomeric repeat amplification, without shortening or lengthening of the telomerase products: A method to analyze the processivity of telomerase enzyme. Nucleic Acids Research. 2001;**29**(2):E3. DOI: 10.1093/nar/29.2.e3

[96] Bazin H, Preaudat M, Trinquet E, Mathis G. Homogeneous time resolved fluorescence resonance energy transfer using rare earth cryptates as a tool for probing molecular interactions in biology. Spectrochimica Acta. Part A, Molecular and Biomolecular Spectroscopy. 2001;**57**(11):2197-2211

[97] Gabourdes M, Bourgine V, Mathis G, Bazin H, Alpha-Bazin B. A homogeneous time-resolved fluorescence detection of telomerase activity. Analytical Biochemistry. 2004; **333**(1):105-113

[98] Savoysky E, Akamatsu K, Tsuchiya M, Yamazaki T. Detection of telomerase activity by combination of TRAP method and scintillation proximity assay (SPA). Nucleic Acids Research. 1996;**24**(6):1175-1176

[99] Hirose M, Abe-Hashimoto J, Ogura K, Tahara H, Ide T, Yoshimura TJ. A rapid, useful and quantitative method to measure telomerase activity by hybridization protection assay connected with a telomeric repeat amplification protocol. Journal of Cancer Research and Clinical Oncology. 1997;**123**(6):337-344

[100] Mayfield MP, Shah T, Flannigan GM, Hamilton Stewart PA, Bibby MC. Telomerase activity in malignant and benign bladder conditions. International Journal of Molecular Medicine. 1998;**1**(5):835-840

[101] Hoos A, Hepp HH, Kaul S, Ahlert T, Bastert G, Wallwiener D. Telomerase activity correlates with tumor aggressiveness and reflects therapy effect in breast cancer. International Journal of Cancer. 1998; **79**(1):8-12

[102] Chen L, Huang J, Meng F, Zhou N. Distinguishing tumor cells via analyzing intracellular telomerase activity. Analytical Sciences. 2010;**26**(5):535-538

[103] Hou M, Xu D, Bjorkholm M, Gruber A. Real-time quantitative telomeric repeat amplification protocol assay for the detection of telomerase activity. Clinical Chemistry. 2001;**47**(3): 519-524

[104] Elmore LW, Forsythe HL, Ferreira-Gonzalez A, Garrett CT, Clark GM, Holt SE. Real-time quantitative analysis of telomerase activity in breast tumor specimens using a highly specific and sensitive fluorescent-based assay. Diagnostic Molecular Pathology. 2002;**11**(3): 177-185

[105] Heller-Uszynska K, Kilian A. Microarray TRAP: A high-throughput assay to quantitate telomerase activity. Biochemical and Biophysical Research Communications. Oct 2004;**323**(2): 465-472

[106] Ohyashiki K, Ohyashiki JH, Nishimaki J, Toyama K, Ebihara Y, Kato H, et al. Cytological detection of telomerase activity using an in situ telomeric repeat amplification protocol assay. Cancer Research. 1997;**57**(11): 2100-2103

[107] Yahata N, Ohyashiki K, Ohyashiki JH, Iwama H, Hayashi S, Ando K, et al. Telomerase activity in lung cancer cells obtained from bronchial washings. Journal of the National Cancer Institute. 1998;**90**(9): 684-690

[108] Ohyashiki K, Yahata N, Ohyashiki JH, Iwama H, Hayashi S, Ando K, et al. A combination of semiquantitative telomerase assay and in-cell telomerase activity measurement using exfoliated urothelial cells for the detection of urothelial neoplasia. Cancer. 1998;**83**(12):2554-2560

[109] Dejmek A, Yahata N, Ohyashiki K, Kakihana M, Hirano T, Kawate N, et al. Correlation between morphology and telomerase activity in cells from exfoliative lung cytologic specimens. Cancer. 2000;**90**(2):117-125

[110] Dejmek A, Yahata N, Ohyashiki K, Ebihara Y, Kakihana M, Hirano T, et al. In situ telomerase activity in pleural effusions: A promising marker for malignancy. Cytopathology. 2001; **24**(1):11-15

[111] Youssef N, Paradis V, Ferlicot S, Bedossa P. In situ detection of telomerase enzymatic activity in human hepatocellular carcinogenesis. The Journal of Pathology. 2001;**194**(4): 459-465

[112] Hirose M, Abe-Hashimoto J, Tahara H, Ide T, Yoshimura T. New method to measure telomerase activity by transcription-mediated amplification and hybridization protection assay. Clinical Chemistry. 1998;**44**:2446-2452

[113] Blackburn EH, Greider CW, Henderson E, Lee MS, Shampay J, Shippen-Lentz D. Recognition and elongation of telomeres by telomerase. Genome. 1989;**31**(2):553-560

[114] Maesawa C, Inaba T, Sato H, Iijima S, Ishida K, Terashima M, et al. A rapid biosensor chip assay for measuring of telomerase activity using surface plasmon resonance. Nucleic Acids Research. 2003;**31**:e4

[115] Rad'ko SP, Voronina SA, Gromov AV, Gnedenko OV, Bodoev NV, Ivanov AS, et al. Use of oligonucleotides conjugated to gold nanoparticles and streptavidin for amplifi cation of optical biosensor signal during detection of telomeric repeats. Bulletin of Experimental Biology and Medicine. 2009;**147**(6):746-749

[116] Grimm J, Perez JM, Josephson L, Weissleder R. Novel nanosensors for rapid analysis of telomerase activity. Cancer Research. 2004;**64**:639-643

[117] Pavlov V, Willner I, Dishon A, Kotler M. Amplified detection of telomerase activity using electrochemical and quartz crystal microbalance measurements. Biosensors & Bioelectronics. 2004;**20**(5):1011-1021

[118] Li Y, Liu B, Li X, Wei Q. Highly sensitive electrochemical detection of human telomerase activity based on bio-barcode method. Biosensors & Bioelectronics. 2010;**25**(11):2543-2547

[119] Schmidt PM, Matthes E, Scheller FW, Bienert M, Lehmann C, Ehrlich A, et al. Real-time determination of telomerase activity in cell extracts

using an optical biosensor. Biological Chemistry. 2002;**383**:1659-1666

[120] Patolsky F, Gill R, Weizmann Y, Mokari T, Banin U, Willner I. Lighting-up the dynamics of telomerization and DNA replication by CdSe-ZnS quantum dots. Journal of the American Chemical Society. 2003;**125**(46):13918-13919

[121] Zheng G, Patolsky F, Cui Y, Wang WU, Lieber CM. Multiplexed electrical detection of cancer markers with nanowire sensor arrays. Nature Biotechnology. 2005;**23**(10):1294-1301

[122] Sharon E, Freeman R, Riskin M, Gil N, Tzfati Y, Willner I. Optical, electrical and surface plasmon resonance methods for detecting telomerase activity. Analytical Chemistry. 2010;**82**:8390-8397

[123] Xu SQ, He M, Yu HP, Wang XY, Tan XL, Lu B, et al. Bioluminescent method for detecting telomerase activity. Clinical Chemistry. 2002;**48**(7): 1016-1020

[124] Zhou X, Xing D, Zhu D, Jia L. Magnetic bead and nanoparticle based electrochemiluminescence amplification assay for direct and sensitive measuring of telomerase activity. Analytical Chemistry. 2009;**81**(1):255-261

[125] Wu JY, Stone MD, Zhuang X. A single-molecule assay for telomerase structure-function analysis. Nucleic Acids Research. 2010;**38**(3):e16. DOI: 10.1093/nar/gkp1033

[126] Mohamad T, Ramakrishnan L, Sachdev HS, Tandon N, Roy A, Bhargava SK, et al. Validation of quantitative polymerase chain reaction with southern blot method for telomere length analysis. Future Science OA. 2018;**4**(4):FSO282

[127] O'Callaghan NJ, Fenech M. A quantitative PCR method for measuring absolute telomere length. Biological Procedures Online. 2011;**13**:3

[128] Kimura M, Stone RC, Hunt SC, et al. Measurement of telomere length by the southern blot analysis of terminal restriction fragment lengths. Nature Protocols. 2010;**5**(9):1596-1607

[129] Cawthon RM. Telomere measurement by quantitative PCR. Nucleic Acids Research. 2002;**30**(10): e47.26

[130] Cawthon RM. Telomere length measurement by a novel monochrome multiplex quantitative PCR method. Nucleic Acids Research. 2009;**37**:e21. Paper describing previous procedure with full details of multiplex qPCR method

[131] Hultdin M, Gronlund E, Norrback K, Eriksson-Lindström E, Just T, Roos G. Telomere analysis by fluorescence in situ hybridization and flow cytometry. Nucleic Acids Research. 1998;**26**(16):3651-3656

[132] Baird DM, Rowson J, Wynford-Thomas D, Kipling D. Extensive allelic variation and ultrashort telomeres in senescent human cells. Nature Genetics. 2003;**33**(2):203-207

[133] De Vivo I, Prescottn J, Wong JY, Kraft P, Hankinson SE, Hunter DJ. A prospective study of relative telomere length and postmenopausal breast cancer risk. Cancer Epidemiology, Biomarkers & Prevention. 2009;**18**(4): 1152-1156. Paper describing previous use of qPCR method and observed higher levels of CVs

[134] Shen J, Terry MB, Gurvich I, Liao Y, Senie RT, Santella RM. Short telomere length and breast cancer risk: A study in sister sets. Cancer Research. 2007;**67**(11):5538-5544. Paper describing previous use of qPCR method and observed higher levels of CVs

[135] Hsieh AYY, Saberi S, Ajay kumar A, et al. Optimization of a relative telomere length assay by monochromatic multiplex real-time quantitative PCR on the light cycler 480: Sources of variability and quality control considerations. The Journal of Molecular Diagnostics. 2016;**18**(3): 425-437

Chapter 2

Telomerase in Space and Time: Regulation of Yeast Telomerase Function at Telomeres and DNA Breaks

Yulia Vasianovich, Alexandra Krallis and Raymund Wellinger

Abstract

A development of new strategies against telomerase-associated disorders, such as dyskeratosis congenita, aplastic anemia or cancer, relies on a detailed understanding of telomerase life cycle and the multiple layers of its regulation. *Saccharomyces cerevisiae* is a prime model to study telomerase function and it has already revealed many conserved pathways for telomerase biology. In this chapter, we review the current knowledge of the regulatory pathways that control telomerase function in budding yeast. In particular, we discuss the cell cycle-dependent assembly of telomerase and its recruitment to telomeres. We also focus on the mechanisms that target telomerase to short telomeres. Finally, we discuss possible pathways that inhibit telomerase function at DNA double-strand breaks, thus limiting deleterious *de novo* telomere addition events.

Keywords: telomerase, cell cycle, regulation, RNP biogenesis

1. Introduction

Eukaryotic chromosomal DNA must be completely duplicated for both daughter cells to receive a full complement of DNA during cell division. Given the inherent properties of the conventional replication machinery, a newly replicated lagging-strand DNA is always slightly shorter than the template parental strand at the ends of chromosomes [1, 2]. As a result, the lagging-strand chromosome end acquires a short 3′-overhang—a conserved feature of ends of linear chromosomes crucial for genome stability. At the same time, the leading-strand ends are initially generated as blunt-ended, and will therefore need to be resected to restore the 3′-overhang [3–5]. This process is repeated during each replication cycle and will inevitably lead to a progressive shortening of the chromosome ends and loss of vital genetic information [6, 7].

In order to meet these end-replication challenges, chromosome ends are capped with a stretch of noncoding DNA repeats, called telomeres. The actual length of these repeat stretches within one cell is not uniform, and slight length variations can occur without any consequences for cell viability and fitness [8]. Moreover, shortened tracts that reach a lower limit of functionality can be restored to a longer form by a specialized reverse transcriptase, called telomerase [9, 10]. This ribonucleoprotein (RNP) enzyme synthesizes new telomeric repeats using its intrinsic protein catalytic subunit and a segment of its internal RNA as a template.

Given its essential task in maintaining genome stability, it is not surprising that telomerase function is tightly controlled at multiple levels. Disruption of this fine-tuned regulation leads to telomerase malfunction or its unregulated expression, which may contribute to serious genetic disorders such as dyskeratosis congenita, aplastic anemia and cancer [11].

Historically, budding yeast has been an extremely rich source of information on the biology of telomeres and telomerase. Therefore, in this review we discuss the intricate network of *Saccharomyces cerevisiae* regulatory mechanisms, that cooperate to ensure timely and sufficient telomerase activity at telomeres. In particular, we focus on the cell cycle-dependent regulation of telomerase assembly and its recruitment to telomeres. We also discuss the mechanisms that target telomerase to short telomeres, which pose a major threat to genome stability, and therefore must be dealt with immediately. Finally, we describe the mechanisms that limit telomerase function at DNA double-strand breaks (DSBs), thereby preventing deleterious *de novo* telomere addition events.

2. Interactions between telomerase components within the RNP complex

In budding yeast, the telomerase holoenzyme consists of the noncoding TLC1 RNA that contains a template for telomere synthesis, and several protein subunits that are bound onto the RNA [8] (**Figure 1A**). The catalytic activity of telomerase relies on the Est2 protein, which together with TLC1 is sufficient for the enzymatic activity of telomerase *in vitro* [12, 13]. The yeast Est2 protein, as well as the catalytic telomerase subunits from other organisms (e.g., human, mouse, fission yeast and ciliates), share sequence homology to the reverse transcriptase family of DNA polymerases, and therefore are collectively known as telomerase reverse transcriptases (TERTs) [12, 14–21]. In addition to Est2, several accessory telomerase subunits associate with the telomerase RNA to mediate telomerase function *in vivo*: Est1 is required for telomerase recruitment to telomeres via its association with the Cdc13 single-stranded telomeric DNA-binding protein; Est3 provides a regulatory function which so far is not very well understood; a set of essential Pop proteins, Pop1, Pop6 and Pop7, stabilizes the association of Est1 and Est2 with TLC1 [22–25]. Elimination of TLC1 or any of the three Est proteins leads to a progressive loss of telomeres and subsequent cellular senescence (an ever shorter telomere, *est*, phenotype), underscoring the importance of all these components for telomerase function *in vivo*. The same *est* phenotype is also observed if the interaction of Cdc13 with Est1 is abrogated, or Pop protein binding to TLC1 is disrupted [22, 25–27]. The TLC1 RNA also associates with the Yku70/80 heterodimer and the Sm_7 complex [28, 29]. In the context of telomerase function, Yku is important for its retention in the nucleus and recruitment to telomeres, while the Sm_7 complex participates in TLC1 maturation and promotes its stability [28, 30, 31].

Several distinct elements have been mapped onto the two-dimensional structure of the TLC1 RNA that was defined by phylogenetic analyses [32] (**Figure 1A**). At the heart of TLC1 lies a group of conserved core elements associated with the reverse transcriptase activity: the single-stranded template, the template boundary element and the pseudo-knot structure, to which the catalytic subunit Est2 is attached [32–34]. The core of the RNA is branched into three stem-loops [32]. A conserved three-way junction brings TLC1 3′- and 5′-ends to the same vicinity and also contains the Sm_7 binding motif. The second stem-loop structure holds the Yku complex [28]. The third arm contains the bulge-stem IVc, which supports Est1 binding at the base [35, 36] and the set of Pop proteins at the distal end [25, 37].

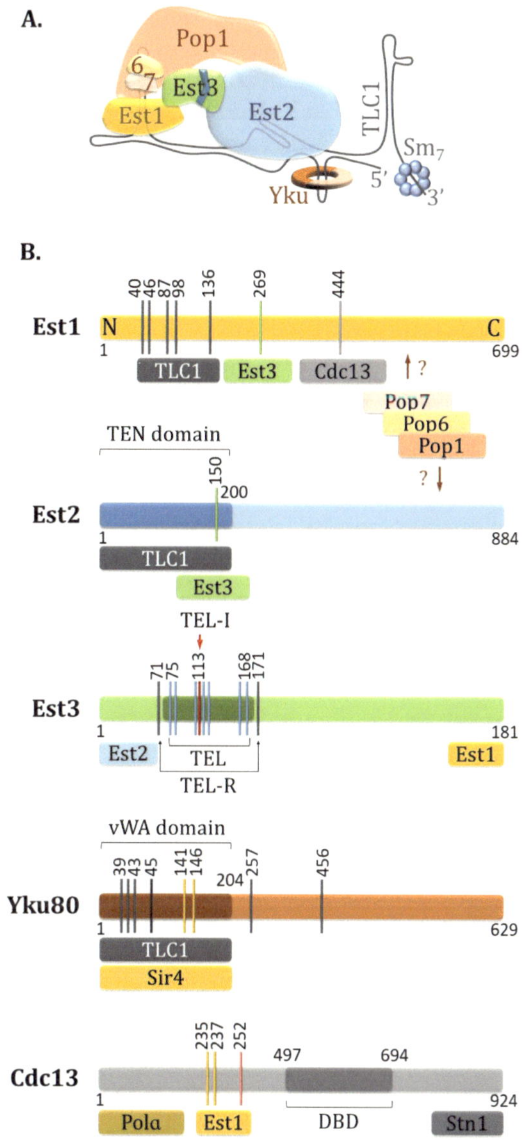

Figure 1.
Telomerase composition and interactions between telomerase components within the RNP. (A) The telomerase RNP consists of the TLC1 RNA, the Est2 catalytic protein and a set of accessory proteins: Est1, Est3, Pop1, 6 and 7, Yku70/80 and the Sm_7 complex (see the text for details). (B) Est1 associates stably with TLC1 in S-phase through the residue cluster at the Est1 N-terminus (dark gray lines). Est3 binds Est1 near the Est1-TLC1 interaction domain in late S-phase. Est1-Est3 binding is abolished by est1-R269E mutation (green line). Cdc13 associates with Est1 in late S-phase. This interaction is compromised in est1-60 (est1-K444E) mutants (light gray line) due to the disruption of a salt bridge between the Est1 and Cdc13 proteins. Pop1, 6 and 7 associate with Est1 and Est2 but the exact interaction surfaces are unknown. Est2 association with TLC1 occurs throughout the cell cycle and is mediated by the Est2 TEN domain. TEN is also required for Est2 binding to telomeric DNA and Est2-Est3 interaction in late S-phase. The Est2-N150D mutant protein (green line) does not bind Est3. Est3 binding to the RNP occurs in late S-phase and requires simultaneous Est3 interaction with Est1 and Est2. The Est3 TEL patch (blue lines) is a cluster of residues that promote Est3 association with telomerase. The Est3 TEL-I patch (Est3-S113Y) (red line) inhibits Est3 binding to the RNP. The Est3 separation-of-function patch (TEL-R) (gray lines) is essential for telomere maintenance but does not affect Est3 interaction with the known telomerase components. Yku80 as part of the Yku70/80 heterodimer associates with TLC1 throughout the cell cycle. This interaction is supported by the vWA domain of Yku80 (gray lines indicate important residues). yku80-135i mutant with a 5 amino acid insertion in the vWA domain (at Yku80-W45, black line) are defective in Yku80-TLC1 interaction. Sir4 also binds to the Yku80 vWA domain. This interaction is strengthened by Yku80-D141 and -E146 residues (yellow lines). Yku80-TLC1 and Yku80-Sir4 interactions can occur simultaneously. Cdc13 binds Est1 via two domains (yellow and pink lines, respectively). cdc13-2 (cdc13-E252K) mutation abolishes the salt bridge between Cdc13 and Est1. Cdc13 binds the 3′-single-stranded telomeric overhang via its DNA binding domain (DBD). Cdc13 also binds Pol1 and Stn1 via its N- and C-terminus, respectively.

A complex hierarchy of the protein subunits defines telomerase composition. Est1 and Est2 bind to separate regions of TLC1 RNA and do not interact directly [33, 34, 38]. Accordingly, Est1 and Est2 do not co-immunoprecipitate in the absence of TLC1 [38]. In addition, Est1 and Est2 binding to TLC1 is not interdependent: disruption of Est1 association with TLC1 does not affect Est2 binding, and vice versa, when Est2-TLC1 interaction is abolished, Est1 still maintains association with TLC1 [34, 35].

In contrast to Est1 and Est2, Est3 does not bind to TLC1 directly. Instead, it associates with telomerase via simultaneous binding to Est1 and Est2, bridging them together. On one end, Est3 interacts with the N-terminus of Est1 (**Figure 1B**). The Est1-Est3 interaction site is distinct from the Est1-TLC1 binding surface, which was also partially mapped to the protein N-terminus [39, 40]. The *est1-R269E* mutation results in the loss of Est1-Est3 interaction and telomere shortening [39, 40]. On the other side, Est3 associates with the TEN domain of Est2 [41–43]. Accordingly, the *est2*^{ten} mutant displays dramatically reduced Est2-Est3 association and telomere shortening [39, 42]. The Est2-Est3 interaction surface overlaps with the Est2 site that binds telomeric DNA, raising a possibility that Est3 might regulate Est2 interaction with telomeres [44]. Notably, a combination of the *est1-R269E* and *est2-N150D* mutations, that attenuate Est3 interaction surfaces in Est1 and Est2, respectively, leads to an additive telomere defect comparable to *est3Δ*. This observation indicates that to function properly, Est3 must be bound to both Est1 and Est2, although only one interaction can be sufficient to support Est3 function [39].

However, it was shown that Est3 cannot interact with Est1 if Est2 is not bound to TLC1, and vice versa, Est3 will not bind Est2 if Est1 is not present in the complex [39]. This suggests that Est3 association with the telomerase RNP absolutely requires both Est1 and Est2 bound to TLC1. Hence, Est3 might have additional interaction surfaces with Est1 and Est2 that are not abolished in *est1-R269E* and *est2*^{ten} mutations. This would explain their milder telomere phenotypes compared to the double mutant.

Interestingly, despite the relative abundance of Est3, only a small fraction of telomerase comprises Est3 during late S-phase, when telomeres are elongated [39, 45, 46]. This argues that Est3 association with the telomerase complex is restricted during the cell cycle. Indeed, an *est3-S113Y* mutation leads to an increased Est3 association with telomerase and elongated telomeres [39]. These data indicate that the area affected by *est3-S113Y* (dubbed TEL-I, for TEL inhibitory) might negatively regulate Est3 recruitment to the telomerase complex (**Figure 1B**). The TEL-I control site seems to affect Est3 binding to both Est1 and Est2, as the *est3-S113Y* mutation partially suppresses the telomere defect caused by Est1 and Est2 mutant proteins that do not interact with Est3.

In addition, Est3 contains a positive regulatory site (named TEL), which promotes Est3 association with Est2 [47]. In contrast to the TEL-I, mutating the TEL patch results in decreased association of Est3 with Est2 and telomere shortening. Notably, both patches are found in close proximity on the Est3 surface. Therefore, it was suggested that TEL and TEL-I might function as a toggle switch, mediating positive and negative regulation of Est3 binding, respectively [39]. Est3 association with the telomerase RNP occurs exclusively during late S-phase of the cell cycle [39, 48]. This is believed to be one of the mechanisms that restrict telomerase function during the cell cycle (discussed in detail in Section 3), thus, raising the possibility that the Est3 toggle switch might be designed to control its cell cycle-specific association with telomerase. Consistent with this idea, disruption of the TEL-I patch results in Est3 binding to telomerase in G1-phase of the cell cycle [39].

Notably, another Est3 regulatory element (the separation-of-function patch, TEL-R) was also identified in proximity to the TEL and TEL-I area [47] (**Figure 1B**).

Although it does not affect Est3 interaction with the telomerase RNP, it is essential for telomere maintenance. It is still unknown what the role of Est3 binding to Est1 and Est2 is; whether it induces a conformational change in telomerase that favors telomere extension or acts as a bridge to attract other yet-to-be-identified telomerase subunits. The existence of the Est3 separation-of-function patch is consistent with both ideas. For instance, it can be a subject of a post-translational modification, which would induce a conformational change in Est3 and affect the rest of telomerase components. Alternatively, it could be the site of a protein-protein interaction with another telomerase component. In both cases, disrupting this patch would abrogate telomere maintenance without affecting Est3 interaction with Est2 or Est1.

Recently, a set of novel telomerase components called Pop1, Pop6 and Pop7 was shown to associate with TLC1 at the P3-like domain of the stem IVc, in the vicinity to the Est1 binding site [25] (**Figure 1A**). These Pop 1/6/7 subunits of telomerase are shared with RNAseP and RNaseMRP, conserved RNP complexes, required for processing of tRNA, rRNA and mRNAs [49]. Notably, within RNAseP and MRP, binding of Pop proteins is also supported by the P3 domain, structurally similar to the one of telomerase. The Pop proteins are essential for telomerase *in vivo* function [25]. Similar to RNAseP/MRP, where Pop1, 6 and 7 stabilize the RNP structure, telomerase Pop subunits are implicated in stabilization of Est1 and Est2 on TLC1 [25, 37]. The Pop-mediated stabilization of telomerase proteins is likely conferred by achieving a precise TLC1 architecture [37]. Introduction of just two base pairs into the short stem between the Est1 and Pop binding sites results in a complete loss of Est1, whereas Pop proteins remained present. This effect was attributed to the rotation of the P3 domain, possibly disrupting interaction between the Pop proteins and Est1. In addition, these results show that the Pop complex binds to telomerase in an Est1-independent manner. Altogether, it suggests that both the functional and physical interaction between the Pop complex and Est1 occurs only in the context of the proper TLC1 architecture, which brings the proteins in close proximity to favor their interaction [37]. Whether the Pop complex also interacts with Est3 and affects its stable association with the telomerase complex, is still unknown and will be important to assess.

The unexpected discovery of the novel Pop telomerase components raises the possibility that the complete telomerase RNP composition is not yet solved and other components are waiting to be uncovered. Interestingly, a distinct set of mutations in the Est1 C-terminus leads to a short telomere phenotype, although all the Est1 interactions known so far remain intact (Est1-TLC1, Est1-Est3, Est1-Cdc13 or Est1-Pop1) [40]. Therefore, Est1 might mediate a novel protein-protein interaction that is important for telomerase function.

3. Regulation of telomerase assembly and disassembly during the cell cycle

In vivo, telomerase function is strictly cell cycle-regulated. As a consequence, telomere elongation only occurs within a narrow window of late S-phase, right after conventional replication is completed [45, 46, 50]. Multiple regulatory mechanisms, discussed throughout this review, ensure that telomerase is active only at the right time and place. One of such mechanisms is the cell cycle-dependent modification of telomerase composition (**Figure 2**). It operates via fine-tuning the abundance of the telomerase protein subunits and their association with the telomerase complex according to the stage of the cell cycle.

The Est2 catalytic subunit and the TLC1 RNA, in essence a minimal telomerase, can in principle associate with telomeres throughout the whole cell cycle [31, 51].

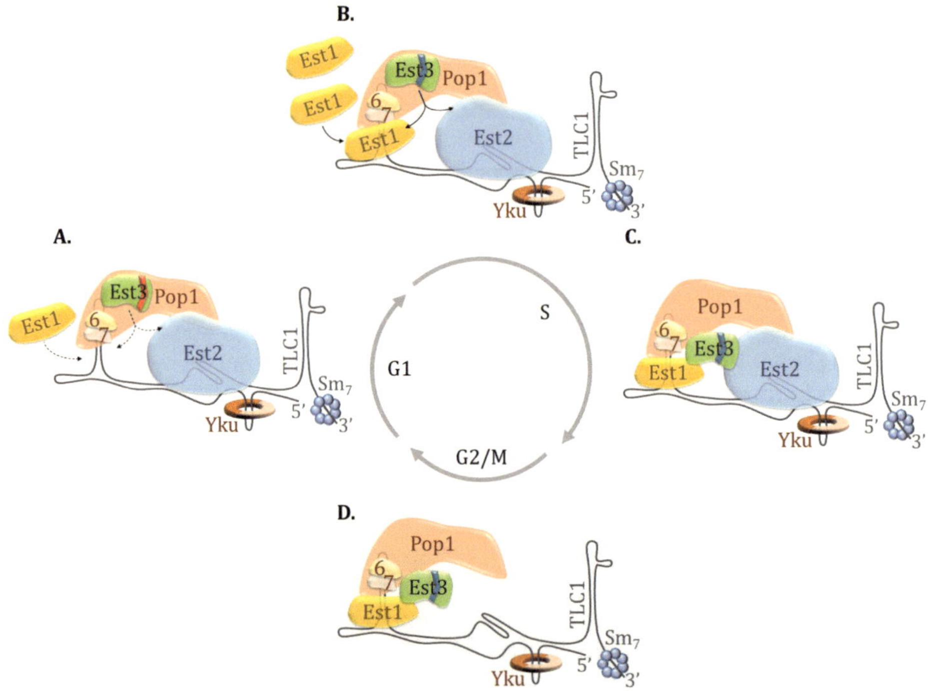

Figure 2.
Telomerase assembly and disassembly through the cell cycle. (A) The Est2 catalytic subunit associates with the TLC1 RNA during G1- and S-phase of the cell cycle. It is likely, that Pop1, 6 and 7 associate with TLC1 constitutively, thus stabilizing the RNA structure and its interactions with other proteins. The Sm_7 complex, required for TLC1 maturation and stability, and the Yku70/80 heterodimer, important for TLC1 retention in the nucleus, are also constitutive subunits of the telomerase RNP. In G1, low abundance of Est1 and activation of Est3 TEL-I patch (red shape) limit their association with the complex. Est3 cannot associate with the telomerase RNP when Est1 is not bound to TLC1. (B) Increased abundance of Est1 in S-phase promotes its association with TLC1. This complex is not active without subsequent Est3 binding promoted by its TEL patch (blue shape) activation. (C) In late S-phase, a functional telomerase complex is formed by Est3 bridging Est1 and Est2. (D) Telomerase is disassembled through the departure of Est2 in G2-phase. Est1 and Est3 remain associated with TLC1, perhaps due to stabilization by Pop1, 6 and 7, or a conformational change induced by initial Est3 association in S-phase.

However, the modes of Est2-TLC1 recruitment to telomeres as well as its consequences vary during different stages. In G1, telomere-telomerase association requires the Yku-TLC1 binding, whereas in late S-phase, it mostly depends on the Cdc13-Est1 interaction. In line with the different mechanisms of telomerase recruitment, association of Est2-TLC1 with telomeres is transient in G1- and G2-phases of the cell cycle. However, during late S-phase it gets stabilized, allowing for productive telomere elongation (see Section 4 for more details) [52].

In contrast to Est2, the total Est1 protein level fluctuates during the progression of the cell cycle (**Figure 2**). While in G1, Est1 abundance is quite low, it reaches its maximum in S-phase and is maintained at the same level for the rest of the cell cycle [48, 51, 53]. This cell cycle regulation of Est1 abundance results in its limited association with telomerase complex and telomeres in G1 [39, 48, 51, 54]. Indeed, Est2 immunoprecipitation experiments revealed that in G1, Est1 association with the telomerase RNP is 3-fold lower than that of Est2 [39, 54]. Only during S-phase does the Est1-Est2 ratio reach 1:1 and remain constant until the end of the cell cycle. Consistent with stable association of Est1 within the telomerase complex in S-phase, Est1 is robustly detected at telomeres at this point of the cell cycle and further on [48, 51]. Notably, by increasing Est1 protein level in G1, the 1:1 ratio of Est1 and Est2 can be achieved in the telomerase complex throughout the cell cycle

[48, 54]. Therefore, Est1 is absent from the telomerase RNP during G1 due to its low abundance, and as of yet there is no evidence for an active exclusion of Est1.

Cell cycle-dependent regulation of Est1 abundance occurs both at the mRNA and protein levels. During G1, the Est1 mRNA level is at its lowest, whereas at the G1-S transition, it increases, reflecting the fact that *EST1* transcription occurs late in G1 [48, 55, 56]. In G2/M-phase, the Est1 mRNA level decreases again, likely as a result of a standard mRNA decay process [55, 56].

Several studies indicate that the low abundance of Est1 in G1-phase is not solely due to cell cycle-specific control of its mRNA level, but also due to the G1-specific proteasome-dependent degradation of Est1 [48, 53, 57]. Degradation by the proteasome requires prior protein poly-ubiquitination [58]. Ferguson et al. showed that Est1 degradation is mediated by the Anaphase Promoting Complex (APC) E3 ubiquitin-ligase, that acts in G1 and ensures a smooth progression of the cell cycle via timely degradation of key regulatory proteins [53]. Disruption of the APC function and APC recognition motifs identified in Est1 was shown to abrogate the cell cycle regulation of the Est1 protein abundance, such that Est1 becomes more stable in G1.

However, this result has been challenged recently, as no change in G1-specific Est1 protein level has been observed in cells bearing mutations in the APC motifs [40]. Ferguson et al. could not detect Est1 poly-ubiquitination *in vivo*, arguing that ubiquitinated proteins would constitute a small fraction of a low abundant Est1 protein, and also would be quickly degraded by the proteasome [53]. However, single mutations of the conserved Est1 lysine residues, which act as the poly-ubiquitination substrates, did not yield a telomere defect, suggesting that Est1 might not be a subject for ubiquitination *in vivo* [40]. The poly-ubiquitination sites could be redundant though, and elimination of the whole lysine cluster might be required to abolish the ubiquitination mechanism. Furthermore, neither APC-dependent degradation, nor poly-ubiquitination of Est1 was observed *in vitro*, suggesting that the APC effect on the Est1 abundance might be indirect [53].

Yet, the idea of the G1-specific degradation of Est1 by the proteasome was supported by another study, which showed that Est1 physically interacts with the Cdc48 complex [57]. Cdc48 is a chaperone, which in complex with the E3 ubiquitin-ligases Npl4 and Ufd1, acts as a segregase to separate ubiquitinated proteins from multi-protein complexes [58]. In cells expressing the *cdc48-3* temperature-sensitive allele at the semi-permissive temperature, the G1-specific regulation of the Est1 abundance was abrogated, and the Est1 level became comparable at all stages of the cell cycle [57]. In addition, the overall Est1 protein level increased almost 40-fold, indicating that Cdc48 contributes to Est1 degradation throughout the cell cycle. Notably, in this study Est1 ubiquitination could be detected *in vivo*. In *cdc48-3* cells, the level of ubiquitinated Est1 was twice higher than in wild-type cells, consistent with the role of Cdc48 in channeling ubiquitinated proteins for degradation. However, enrichment of the ubiquitinated Est1 was observed throughout the whole cell cycle, which again suggests that the Cdc48 role in Est1 degradation is not limited to G1-stage of the cell cycle. It is plausible, that in addition to its regulatory G1 function, the Cdc48 complex promotes Est1 degradation after the disassembly of the telomerase complex in G2 (see below). However, the Est1 protein level does not decrease in G2-phase [48, 51]. This argues that accumulation of the ubiquitinated Est1 population in *cdc48-3* cells is not linked to the Est1 degradation defect, but occurs for another reason [57]. Indeed, although the proteasome-mediated degradation requires protein poly-ubiquitination, only mono-ubiquitinated Est1 was observed in *cdc48-3* cells.

Altogether, these conflicting results leave open the question about the role of the proteasome in cell-cycle regulation of Est1 abundance and require additional clarification [40, 48, 53, 57].

The Est3 protein level *per se* is not the subject of cell cycle regulation. However, in line with the Est1 behavior, Est3 association with telomerase is also cell-cycle regulated, being restricted to late S-phase [39, 48]. Similar behavior of Est1 and Est3 can at least partly be explained by the fact that Est1 is required for Est3 association with Est2 and TLC1 [23, 48] (**Figure 2**).

Notably, overexpression of Est1 in G1 leads to its association with the telomerase RNP, as well as Est3 binding, resulting in the assembly of the active telomerase [48, 54]. Despite that, no productive telomere elongation was observed during G1-stage of the cell cycle [46, 48, 52, 54]. This observation strongly suggests that the cell cycle regulated telomerase assembly is not the limiting regulatory mechanism that prevents telomere elongation during G1 (see Section 4).

As outlined above, the assembly of the telomerase complex is a tightly regulated process that occurs via regulation of abundance and inclusion of the Est1 and Est3 accessory proteins in the complex in late S-phase [39, 48, 51, 54]. However, there is evidence that telomerase disassembly may also be an actively regulated process. It may occur via a mechanism different from telomerase assembly, as it requires the dissociation of the Est2 catalytic subunit, resulting in the formation of the Est1-Est3-TLC1 disassembly complex [39] (**Figure 2D**). It is not known what serves as a signal for Est2 dissociation and how this process occurs mechanistically. Since Est3 interacts with Est2 in the TEN domain, which is also responsible for Est2 binding to the telomeric DNA, it is possible that the G2-specific disassembly of telomerase complex might be signaled via the Est3 subunit [39, 44].

One of the intriguing questions is how Est3 remains within the Est1-Est3-TLC1 complex after Est2 dissociation, given that its binding to the complex requires a simultaneous presence of Est1 and Est2 on TLC1 [39]. It is possible that once bound to the telomerase complex, Est3 can be stabilized by other proteins and consequently no longer requires Est2 to keep its position. One of the candidates for such a function is the complex of the Pop1, Pop6 and Pop7 proteins, which might stabilize Est3 binding within the telomerase complex, as they do for Est1 and Est2 [25, 37]. In general, although constitutively present in the telomerase RNP, the role of the Pop proteins in telomerase regulation remains unexplored. Therefore, it would be of a great interest to assess whether their telomerase-related function is somehow regulated during cell cycle.

4. Regulation of telomerase recruitment to telomeres during the cell cycle

As mentioned above, telomerase assembly is highly regulated during the cell cycle. However, this control is insufficient to limit telomerase function to late S-phase. In this section, we discuss how the telomerase RNP is recruited to telomeres, and how regulation of this process makes an impact on cell-cycle restriction of telomerase function.

4.1 Sir4-Yku-TLC1 as the G1-specific telomerase recruitment mechanism

Telomerase is recruited to telomeres via two different mechanisms: (1) via the Sir4-Yku70/80-TLC1 interaction, which mostly operates in G1-phase of the cell cycle, and (2) via the Cdc13-Est1 interaction, which is the predominant recruitment pathway during late S-phase, when telomere elongation takes place (**Figure 3**).

The Yku complex is not a dedicated telomeric protein, and its main function is normally associated with DSB repair via non-homologous end joining (NHEJ) [59]. Yku is a heterodimer that consists of the Yku70 and Yku80 subunits. Together the

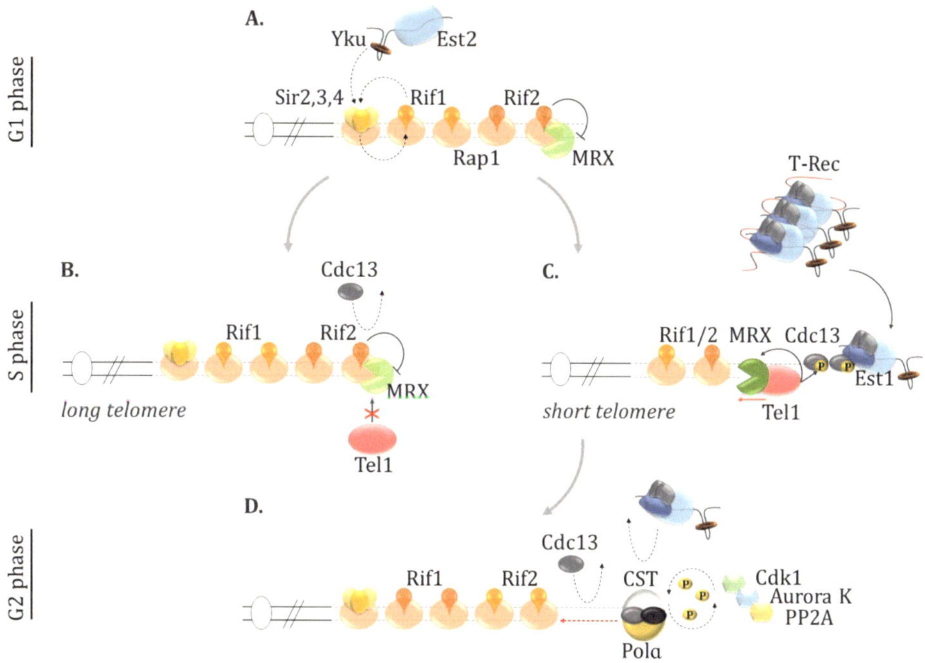

Figure 3.
Recruitment and disassociation of telomerase at short and long telomeres. (A) Inhibition of telomerase in G1: Rap1-bound Rif2 inhibits MRX-dependent telomeric C-strand resection. In addition, Rif1 and Rif2 may compete with Sir proteins for Rap1 binding, resulting in unstable telomerase recruitment via the Sir4-Yku-TLC1 pathway. (B) Inhibition of telomerase at long telomeres in S-phase: Long telomeres allow increased Rap1 binding and recruitment of Rif and Sir complexes. Rif2 competes with Tel1 to inhibit MRX activation, and subsequent 5′-end resection. This prevents the formation of a single-stranded substrate for Cdc13 and telomerase recruitment. (C) Telomerase recruitment at short telomeres in S-phase: At short telomeres, less Rap1 and Rif binding allows TERRA transcription (red line). TERRA recruits multiple telomerase molecules (T-Rec) to its telomere of origin. Tel1 can outcompete Rif2 for MRX binding, promoting 5′-end resection. Cdc13 can bind single-stranded DNA and is phosphorylated by Tel1 to recruit telomerase via Est1. (D) Elongation of telomeres by telomerase is terminated in G2-phase: Cdk1-, Aurora K- and PP2A-mediated phosphorylation and de-phosphorylation of Cdc13 and Stn1 allow formation of the CST (Cdc13, Stn1, Ten1) complex. This inhibits Est1-Cdc13 interaction and therefore decreases telomerase recruitment. Cdc13 and Stn1 recruit DNA polymerase α to fill-in the telomeric C-strand. After the C-strand has been filled-in, Cdc13 no longer has a single-stranded substrate to bind to, and hence cannot recruit telomerase.

two subunits form a ring structure, that binds a DSB end by encircling DNA strands [60]. Despite the fact that NHEJ must be avoided at all costs at telomeres, Yku is present at chromosome ends, where it plays multiple roles, such as inhibition of 5′-telomere processing [61, 62], maintenance of the telomere position effect [61, 63, 64], as well as telomere clustering and tethering to the nuclear envelope [65]. In addition, Yku is required for telomerase retention in the nucleus and its recruitment to telomeres in G1 [30, 31].

Yku70/80 associates with TLC1 via a 48 nt RNA stem-loop [30, 66] (**Figure 1**). Indeed, in *tlc1-Δ48* and *yku80-135i* mutants with abolished Yku-TLC1 interaction, the association of telomerase components with telomeres follows the same pattern as in *yku80Δ* cells. The structural element of TLC1 required for interaction with the Yku complex has been recently narrowed down to 25 nt [67]. It consists of two stems separated by a two-nucleotide (A292U293) bulge. Bending the TLC1 stem-loop in a bulge region is absolutely essential for Yku binding *in vitro* and *in vivo*. Indeed, *tlc1ΔAU* cells have slightly shortened telomeres, comparable to the effect of *tlc1Δ48* with the whole stem-loop being removed.

Structural studies indicate that Yku binds double-stranded DNA and the TLC1 stem-loop through the same aperture in the Yku70/80 heterodimer structure, indicating that Yku cannot simultaneously interact with DNA and RNA [67, 68]. This suggests that interaction of the Yku-TLC1 complex with telomeres might be mediated by a protein-protein interaction. Indeed, Yku80 directly interacts with Sir4, a component of telomeric chromatin that is recruited to telomeres via the Rap1 double-stranded telomere binding protein [69, 70]. The Sir4-Yku80 interaction site is not in conflict with the Yku70/80-TLC1 binding surface [67]. Accordingly, the Yku70/80 heterodimer, TLC1 and Sir4 can be co-immunoprecipitated as a complex, implying that telomerase could be recruited to DNA via Sir4-Yku80 interaction. Indeed, in *sir4* mutants with abolished Sir4-Yku80 interaction, the Est2 level at telomeres is 15-25% lower than in wild-type cells, which is reminiscent of the *sir4Δ* effect. In addition, telomere length in these mutants is somewhat shorter than in wild-type cells, similar to the *yku80* mutants with an attenuated Sir4-Yku80 interaction.

Interestingly, the telomere defect in the *yku80Δ* cells is much more pronounced than in mutants with the disrupted Sir4-Yku80 interaction. This effect can be explained by the additional role of the Yku complex in the nuclear retention of TLC1, which does not depend on Sir4 [31, 67]. Accordingly, the level of nuclear TLC1 was reduced both in the *tlc1ΔAU* and *yku80Δ* cells with abolished Yku-TLC1 interaction [67]. At the same time, in the Sir4-Yku80 interaction mutants, the nuclear fraction of TLC1 was normal.

Certain evidence suggests that at telomeres, the Yku complex does not exist as a uniform population. Apart from the Sir4-bound fraction, Yku also binds to telomeres directly, as it does at DSBs [71–73]. In addition to having two different modes of telomere binding, Yku was also found at diverse telomeric loci, including the junction between telomere and subtelomere regions, as well as interstitial telomeric sequences between the subtelomeric repeats [73]. Both populations include Yku directly bound to DNA and Sir4-bound Yku. It is still unknown whether Yku binds to the very tips of telomeres, and if so, in what fashion. The position of Yku so far from the telomere end was attributed to its role in maintenance of stalled and broken replication forks. Here, Yku might load on one-ended breaks and protect them from end-resection, thus favoring telomere addition by telomerase. However, since the Sir4-bound Yku population is also present at the distal telomeric loci, it suggests that telomerase recruited to telomeres via the Sir4-Yku80 pathway might be kept far from the telomere end. By extension, since the Sir4-Yku80 pathway operates mainly in G1-phase, keeping telomerase at a larger distance from the chromosome end might be a novel regulatory step, which would ensure telomerase molecules are not engaged in productive telomere elongation in G1.

Disrupting the Sir4-Yku-TLC1 pathway of telomerase recruitment results only in mild telomere shortening [28, 66, 67]. This implies that the productive telomerase recruitment required for telomere elongation is predominantly mediated by a different mechanism. Indeed, the S-phase specific Cdc13-Est1 interaction underlies the main functional telomerase recruitment pathway, and is therefore absolutely essential for telomere maintenance [22, 26, 27].

4.2 Cdc13-Est1 as the major telomerase recruitment pathway in late S-phase

A significant feature of late S-phase telomeres that distinguishes them from G1 and G2 chromosome ends is the formation of detectable 3′-single-stranded telomeric overhangs, or G-tails [74, 75]. The formation of telomeric overhangs requires MRX-dependent 5′-end processing, but, as briefly mentioned in the introduction, this process seems to take place only on the leading-strand telomeres [4, 76–78]. This S-phase specific structure is bound by Cdc13 and facilitates telomerase recruitment to telomeres via a direct association between Cdc13 and Est1 [22] (**Figure 3C**).

The Cdc13-Est1 interaction is also favored by an increased abundance of Est1 in late S-phase (see Section 3) [51].

The direct interaction between Cdc13 and Est1 was inferred from the fact that the C-terminal Cdc13 Glu252 residue forms a salt bridge with the Lys444 of Est1 (**Figure 1B**). Abrogation of either of these partners by the *cdc13-2* or *est1-60* mutations, respectively, results in a loss of telomerase recruitment and the senescence phenotype [26, 27]. However, a combination of mutations that leads to a charge swap in those residues, suppresses the defect associated with single mutations [27]. Consistently, co-immunoprecipitation of Est1 via Cdc13 revealed that in *cdc13-2* cells, Cdc13-Est1 binding is significantly reduced, but a complementary *est1-60* mutation can restore this interaction [40, 54].

However, the importance of the Cdc13^{Glu252}-Est1^{Lys444} salt bridge in telomerase recruitment was challenged by several later studies. For example, no interaction defect between the Cdc13-2 mutant protein and Est1 has been detected in the yeast two-hybrid and co-immunoprecipitation analyses [79]. In addition, the *cdc13-2* and *est1-60* mutations only mildly affected Cdc13-Est1 association *in vitro* [67, 80]. Furthermore, estimation of the Est1 level at telomeres yielded the opposite results, reporting either no change or a complete loss of Est1 association with telomeres in *cdc13-2* cells [51, 67, 81]. In any case, neither *cdc13-2* nor *est1-60* mutants can maintain telomeres and the cells undergo senescence, suggesting that an important functional interaction is abrogated by these mutations and can be rescued by the charge swap.

Recent studies shed some light on the *cdc13-2/est1-60* controversy, demonstrating that this specific Cdc13-Est1 interface might be important for activation or stabilization of telomerase after it is recruitment to telomeres [37, 67]. Cells expressing the Cdc13-Est2 fusion bypass the need for Est1 for telomerase recruitment to telomeres [22]. Due to the loss of the Est1-dependent regulatory mechanism of telomerase activity, these mutants acquire over-elongated telomeres. Notably, the long telomere phenotype of the Cdc13-Est2 cells was even further exacerbated if Est1 was ectopically expressed, indicating that Est1 indeed activates or stabilizes telomerase after its recruitment to telomeres [37, 67]. However, expression of the TLC1 variants deficient in Est1 binding in this strain background suppressed the telomere-elongation phenotype, resulting in the wild-type telomere length [37]. Furthermore, supplementing cells expressing the Cdc13-Est2 protein fusion with the *est1-60* allele resulted in telomere shortening when compared to isogenic strains expressing the wild-type Est1 protein [67]. A similar effect was observed when the Cdc13-2 mutant protein was fused with Est2. However, when the Cdc13-2-Est2 fusion was combined with the Est1-60 protein expression, telomere length was restored to that of a wild-type. This result indicates that Est1 association with the telomerase RNP and the interface between the Cdc13 and Est1 forming the Cdc13^{Glu252}-Est1^{Lys444} salt bridge might be required for telomerase activation and/or stabilization, which comes into play downstream of telomerase recruitment [37, 67].

If the Cdc13^{Glu252}-Est1^{Lys444} salt bridge area is important for telomerase stabilization, which Cdc13 interface supports the Est1 recruitment? Based on *in vitro* studies, the Est1-recruitment site of Cdc13 may be located within the N-terminal part of the protein, in close proximity to the salt bridge interface discussed above [67]. Disruption of this area in *cdc13-P235A* and *cdc13-F237A* mutants resulted in a large reduction of telomere length. In addition, *cdc13-F237A* cells displayed an S-phase specific decrease in Est1 and Est2 association with telomeres, whereas the level of Cdc13 was not affected. It must be noted that when compared to *cdc13-2* mutants, the *cdc13-F237A* defects both in telomere maintenance and telomerase binding to telomeres were somewhat weaker [67, 81]. However, based on the assigned functions, the recruitment-deficient *cdc13-F237A* mutant would be expected to have more pronounced telomere phenotypes. Hence, additional studies might be required to dissect the dual role of the Cdc13-Est1 interaction in supporting telomerase recruitment to telomeres and its subsequent stabilization and/or activation.

Certain evidence suggests that Cdc13 is phosphorylated by the Tel1/Mec1 checkpoint kinases to promote the Cdc13-Est1 interaction [82, 83] (**Figure 3C**). Indeed, *in vitro*, Tel1 phosphorylates Cdc13 at the positions 225, 249 and 255 [82]. Mec1/Tel1-dependent Cdc13 phosphorylation was also detected *in vivo* by the S249/S255 phospho-specific antibody [83]. The *cdc13-S249A/255A* mutants with abolished Tel1 phosphorylation sites display telomere shortening and the senescence phenotype, as well as a loss of telomerase-mediated telomere addition at HO sites flanked by telomeric repeats. Notably, the identified phosphorylation sites reside in the part of the Cdc13 protein, which is responsible for the formation of the $Cdc13^{Glu252}$-$Est1^{Lys444}$ salt bridge. This implies that Tel1-mediated Cdc13 phosphorylation might be required for telomerase activation/stabilization after telomerase recruitment. Consistent with this idea, lack of Cdc13 could be complemented by expression of Cdc13-Est1 as well as Cdc13-S249A/255A-Est1 fusion proteins [82]. This suggests that Cdc13 phosphorylation may promote the Est1 function downstream of telomerase recruitment.

However, there are also conflicting data on this issue. In particular, cells expressing the Cdc13 protein in which all potential consensus Tel1 phosphorylation sites were mutated, did not display any telomere defect [84]. In addition, the mass-spectrometry analysis of the Cdc13 phosphorylation sites did not detect *in vivo* phosphorylation of the S249/255 residues [85]. Therefore, the existence of Tel1-mediated Cdc13 phosphorylation *in vivo* and its role in telomerase function remains unclear.

Altogether, the data presented above demonstrates that in principle, telomerase can be recruited to telomeres via two separate mechanisms: Sir4-Yku-TLC1 which operates in G1 and Cdc13-Est1 available in late S-phase. However, live-cell imaging of TLC1 dynamics revealed that in G1 and G2 phases, telomerase-telomere interactions are very short-lived and transient [52]. Indeed, outside S-phase, TLC1 molecules move much more rapidly and diffusively as compared to telomeres. During late S-phase, however, TLC1 RNA molecules can assemble in a cluster of 6-15 molecules, named telomerase-recruitment cluster, or T-Rec, which stably associates with telomeres and follows their dynamics.

Consistent with the *in vivo* pattern of telomerase-telomere interactions, the loss of the Sir4-Yku telomerase recruitment pathway has only a mild effect on telomere length [28, 66, 67], whereas disruption of the Cdc13-Est1 interaction completely abrogates telomere maintenance and results in cellular senescence [22, 26, 27]. Therefore, it can be concluded that the Cdc13-Est1 mechanism is the only true telomerase recruitment pathway which allows efficient telomere elongation. As telomerase recruitment via Cdc13-Est1 is only possible in late S-phase, Cdc13-Est1 interaction might be the main factor that determines cell cycle restriction of telomerase function [51]. As for the Sir4-Yku80 telomerase recruitment pathway, the question about its functional significance remains open (see Section 6 for discussion).

4.3 The switch between the Cdc13-Est1 and Cdc13-Stn1-Ten1 complexes as a signal for telomerase dissociation

In addition to the Cdc13-Est1 interaction important for telomerase recruitment to telomeres, Cdc13 also forms a complex with the Stn1 and Ten1 proteins (CST) (**Figure 3D**). CST prevents degradation of the chromosome ends and their recognition by DSB repair mechanisms, collectively known as the capping function [8]. The CST complex also negatively regulates telomere maintenance, demonstrated by the fact that disrupting Cdc13-Stn1-Ten1 interactions leads to a long telomere phenotype [86]. The CST-dependent effect on telomere elongation is due to a direct inhibition of telomerase recruitment, as well as an indirect consequence of the CST function in the re-synthesis of the telomeric C-strand, which eliminates the G-tails as a substrate for telomerase recruitment [86, 87].

The idea of CST directly inhibiting telomerase recruitment stems from two-hybrid experiments, indicating that Stn1 and Est1 compete for binding to Cdc13 [87]. Indeed, the mutant allele *cdc13-2*, that disrupts the Cdc13-Est1 interaction, also abolishes the Cdc13-Stn1 binding, suggesting that Est1 and Stn1 might share the same Cdc13-binding interface. Consistently, the senescence phenotype of the *cdc13-2* mutants is partially suppressed by Est1 overexpression, while overexpression of Stn1 enhances it. To support the idea of the Est1-Stn1 competition, an interesting experiment was performed with mutants expressing only the N-terminus of Stn1, which does not bind Cdc13 [86]. As in other strains where the Cdc13-Stn1 interaction is disrupted, these cells harbored long telomeres. However, an additional expression of the C-terminal portion of Stn1 suppressed this phenotype. Evidently, the C-terminus, expressed as a separate protein unit, cannot simply attach itself to the rest of Stn1, and thus, mediate the Cdc13-Stn1 interaction. The more likely explanation would be that the Stn1 C-terminus interacts with Cdc13 independently of the rest of Stn1 to outcompete Est1 for Cdc13 binding. Hence, Est1 and Stn1 compete for the same Cdc13 interface, implying that the formation of the CST complex can directly inhibit telomerase recruitment.

Telomerase-mediated elongation of the G-strand is followed by C-strand fill-in synthesis, which restores the double-stranded portion of the telomere (**Figure 3D**). CST plays an essential role in this process, as it is responsible for the recruitment of the DNA polymerase α/primase complex to telomeres. For example, it has been shown that Cdc13 interacts with Pol1, the catalytic subunit of this complex [79], while Pol12, the B subunit of the complex, associates with Stn1 [86, 88]. As a consequence of C-strand synthesis, the single-stranded telomeric overhangs that serve as substrates for Cdc13 binding are eliminated. Hence, this effect contributes to the CST role as a negative regulator of telomerase recruitment. Consistently, cells deficient in the Cdc13-Pol1 interaction have elongated telomeres, and this phenotype is dependent on telomerase interaction with Cdc13 [87]. Surprisingly, cells expressing Cdc13 with an N-terminal truncation are proficient in Pol1 interaction, but still have a telomere elongation phenotype as well as longer telomeric overhangs. The latter phenotype was attributed to the defect in the C-strand fill-in, as it was telomerase-dependent, the overhangs were only visible during late S-phase, and the C-strand was also slightly elongated when compared to wild-type cells. Overexpression of Stn1 almost completely suppressed the long telomere and G-tail phenotype in these mutants. This suppression mechanism might be mediated by the Stn1-Pol12 interaction, which could stabilize DNA polymerase α at telomeres. Consistently, certain *pol12* alleles also confer telomere elongation and longer single-stranded telomeric overhangs [88].

Altogether, the above results suggest that the CST complex can affect telomerase recruitment via two mechanisms: (1) directly, via competition between Stn1 and Est1 for Cdc13 binding; and (2) indirectly, via its role in the restoration of the double-stranded telomere structure and elimination of the G-tails. If the Cdc13-Est1 complex permits telomerase recruitment, and the CST complex prohibits it, the switch between these two complexes might serve as an additional regulatory mechanism, which determines the window of telomerase function at telomeres.

Indeed, the transition from the Cdc13-Est1 interaction to the CST complex formation seems to be regulated in part by sequential phosphorylation of Cdc13 and Stn1 by Cdk1 [89, 90] (**Figure 3D**). Cdk1-dependent phosphorylation of Cdc13 and Stn1 is mediated by S- and M-phase cyclins, respectively [90]. During S-phase, Cdc13 phosphorylation by Cdk1 leads to its association to telomeres, where Cdc13 can form a complex with Est1 and recruit telomerase. Moreover, S-phase cyclin-dependent Cdc13 phosphorylation also delays the phosphorylation of Stn1 by M-phase cyclins. In turn, Cdk1-dependent Stn1 phosphorylation promotes the formation of the CST complex.

In addition, Tel1/Mec1 checkpoint kinases also contribute to the balance between the Cdc13-containing telomeric complexes. As was mentioned above, Tel1/Mec1-mediated Cdc13 phosphorylation promotes the Cdc13-Est1 interaction [82, 83]. In turn, this effect is counteracted by the PP2A phosphatase and the Aurora kinase [83]. Pph22, the catalytic subunit of the PP2A phosphatase, dephosphorylates Cdc13 sites phosphorylated by Tel1 or Cdk1. This facilitates telomerase dissociation from telomeres in G2/M-phase, which is demonstrated by prolonged Est1 and Est2 telomere association in *pph22Δ* strains. Aurora kinase on the other hand, phosphorylates Cdc13 in G2/M, thus promoting TLC1 departure from telomeres, possibly by allosterically influencing its interactions with Est1.

Therefore, a tightly regulated sequence of Cdc13 and Stn1 phosphorylation events, first, promotes the formation of the Cdc13-Est1 complex, and next, mediates the shift to the Cdc13-Stn1-Ten1 complex. Such a precise order of events defines a narrow time frame, which permits telomerase function at telomeres, explaining how the cell-cycle regulated restriction of telomerase function is orchestrated.

5. Regulation of telomerase preference for short telomeres

5.1 Rif2-MRX-Tel1 mechanism

Telomerase function is restricted not only by the cell cycle, but also by telomere length, being preferentially targeted to short telomeres. Indeed, within a particular cell cycle, only 6–8% of wild-type length telomeres (~300 bp) get extended, whereas short telomeres of about 100 bp are elongated in almost 45% cases [91]. Such preference for short telomeres is mediated by the Rap1-interacting partners Rif1 and Rif2. Altogether, Rap1, Rif1, and Rif2 form a negative feedback loop, which regulates telomere elongation in a length-dependent manner [92, 93] (**Figure 3**). Therefore, telomeres become over-elongated in the absence of Rif1 or Rif2, due to the increased frequency of telomerase function at all telomeres [91, 94–96]. The Rap1-Rif1-Rif2 regulatory mechanism relies on the number of the telomere-associated Rif proteins as a readout of an individual telomere length [93]. As a result of such "protein counting," only those telomeres that have the low number of Rif1 and Rif2, i.e., short ones, will be elongated. Once telomeres get extended and the sufficient amount of the Rif proteins is restored at telomeres, the negative feedback loop inhibits telomere extension.

The mechanism of the Rif protein counting, and hence, targeting telomerase to short telomeres, depends on the intricate network of physical and functional interactions between the Rif1 and Rif2 proteins, Tel1 and the MRX (Mre11-Rad50-Xrs2) complex.

The checkpoint kinase Tel1 preferentially localizes to short telomeres, and as a result, also mediates Est1 and Est2 preference to short telomeres [97–100]. Tethering Rif1 and Rif2 to DSB ends leads to the reduction in Tel1 binding, suggesting that Rif proteins might out-compete or displace Tel1 from chromosome ends. Indeed, both Rif2 and Tel1 are recruited to telomeres via the Xrs2 subunit of the MRX complex, whereas Rif1 recruitment is partially mediated by Rif2 [96, 101]. *In vitro,* the Tel1-Xrs2 interaction is inhibited by Rif2, suggesting that Rif2 and Tel1 compete for Xrs2 binding. However, it is unclear whether Rif2 and Tel1 binding to Xrs2 is mutually exclusive. Alternatively, binding of Rif2 may stimulate Tel1 dissociation [96].

As was mentioned above, timely generation of the single-stranded 3′-telomeric overhangs is crucial for telomerase function. This process largely relies on the MRX/Sae2 complex, which resects the telomeric 5′-strand and generates a short G-tail in late S-phase after the passage of the replication fork [50, 76, 77]. In addition, the Mre11-Sae2 complex is essential for opening the hairpin structures that might be formed at telomeres by G-strand foldbacks, followed by DNA synthesis and hairpin closure [102]. If not resolved, replication of such hairpins will result in the formation of dicentric chromosomes and initiation of a breakage-fusion-bridge cycle that is detrimental for genome stability.

It was shown, that a balance of the Tel1 and Rif2 activities defines the extent of telomere processing by MRX [103–105]. On one hand, Tel1 enhances MRX-dependent 5′-telomere processing [103]. Rif2, on the other hand, inhibits MRX activity [104, 105]. Tel1 also increases association of MRX at the breaks flanked with telomeric repeats, although it is unclear, if this effect also exists at native telomeres [96].

Based on these observations, the following model for the regulation of telomerase preference for short telomeres by the Rap1-Rif1-Rif2 counting mechanism can be proposed. At short telomeres, Rif2 levels are reduced [99], which relieves inhibition of MRX-mediated telomere processing and also allows unrestricted Tel1 interaction with MRX. In turn, Tel1 may stabilize MRX association with telomeres and enhance its resection activity. As a result, short telomeres acquire single-stranded G-overhangs that serve as a substrate for Cdc13 binding and subsequent Est1-mediated recruitment of telomerase [84, 103]. Once telomeres are extended, the double-stranded Rap1-binding sites are restored. As a result, more abundant Rif2 binding to telomeres decreases Tel1 association with MRX and inhibits MRX-dependent resection, thus blocking Cdc13 and telomerase access to telomeres. Hence, the availability of the G-tail as a substrate for telomerase binding is the main feature that distinguishes short telomeres from the long ones and allows their preferential elongation.

As an alternative to the Rap1-Rif1-Rif2 protein-counting mechanism described earlier, the telomerase preference for short telomeres could also be explained by a replication fork model, proposed recently by Carol Greider [106]. This model predicts that telomerase associates and progresses with the conventional replication fork, and telomere elongation can happen only if telomerase successfully reaches the chromosome end. The telomeric proteins (such as Rap1-Rif1-Rif2) may pose an obstacle for telomerase movement with the replication fork. Hence, the longer the telomere, the longer the distance telomerase must cover to reach the telomere, and the higher the chance for its premature dissociation. At short telomeres, the probability that telomerase will reach the chromosome end increases, explaining why short telomeres are preferentially extended in a given cell cycle. The late timing of telomere replication also justifies the cell cycle-restricted mode of the telomerase function, which also occurs in late S-phase.

However, in direct conflict to the replication fork model, Gallardo et al. showed that in the absence of Rif1 and Rif1 telomeric proteins, telomerase can elongate telomeres in G1-phase [52]. Since, replication does not take place in G1, these data show that a functional telomere-telomerase interaction as well as telomere elongation by telomerase do not depend on the replication fork. However, it is possible, that the replication fork might operate as an alternative pathway, which functions under specific circumstances.

5.2 TERRA-mediated pathway

Another mechanism promoting telomerase recruitment to short telomeres depends on the long noncoding RNA, TERRA (telomeric repeat-containing RNA), transcribed from telomeres [107, 108]. In yeast, TERRA is preferentially expressed from short telomeres and reaches its maximum level in S-phase [109, 110]. During late S-phase, TERRA may act as a scaffold to allow the formation of a cluster of telomerase molecules (T-Rec), and promote telomerase recruitment to the short telomere of the TERRA origin [52, 109] (**Figure 3C**). Indeed, FISH experiments revealed that TERRA interacts with TLC1 RNA, and the complex of TERRA and T-Rec co-localize with a short telomere, from which TERRA is expressed [109].

Interestingly, TERRA expression is inhibited by Rif1 and Rif2 [111]. Therefore, it is possible that at short telomeres, containing fewer Rap1-binding sites and less Rif proteins, TERRA expression might be de-repressed, in order to favor recruitment of telomerase to short telomeres [109, 111]. Hence, the Rif counting mechanism, which allows preferential extension of short telomeres, may operate not only via Tel1-MRX-dependent formation of G-tails, but also via TERRA expression from short telomeres [91, 96, 103, 109].

6. Why telomerase is not welcome in G1-phase

As was discussed in Section 3, Est1 and Est3 do not associate stably with the telomerase RNP in G1-phase [39, 48, 51, 54]. However, restoration of telomerase composition in G1, still did not result in telomere extension [48, 54]. This implies that lack of telomerase components is not the main factor that restricts telomerase function in G1. Instead, the Rif1 and Rif2 telomeric proteins might govern such a regulatory mechanism.

Indeed, in the absence of Rif1 or Rif2, telomerase clusters are no longer restricted to late S-phase of the cell cycle and also appear in G1 [52]. Moreover, these G1-born T-Recs are functional in *rif1Δ* or *rif2Δ* mutants, as they can efficiently elongate short telomeres.

The most plausible explanation for this phenomenon lies in the role of Rif proteins in inhibition of telomere processing and formation of 3′-single-stranded overhangs [103, 104] (**Figure 3A**). Indeed, Rif2 was shown to inhibit processing in both G1 and G2 [104]. This means that in *rif2Δ* mutants, G-tails might be generated outside S-phase, followed by Cdc13-dependent telomerase recruitment. In addition, unrestricted expression and accumulation of TERRA in *rif1Δ* or *rif2Δ* mutants may contribute to unconstrained telomerase function, by promoting formation of telomerase clusters at any stage of the cell cycle [109].

Alternatively, telomere elongation may not take place in G1 due to unstable association of telomerase with telomeres outside S-phase [52]. Rif1 and Rif2 are recruited to telomeres via interaction with Rap1, and hence, compete with Sir3 and Sir4 which also bind to Rap1 [95, 112]. As a result of this competition, Sir4-Yku binding to telomeres might be unstable, providing an explanation for transient telomerase recruitment to telomeres in G1 (**Figure 3A**). Indeed, the role of Rif proteins as negative regulators of telomere length affects the Sir4-Yku pathway of telomere maintenance [70]. *rif1Δ* and *rif2Δ* mutants have hyper-elongated telomeres [94, 95]. The *tlc1Δ48* mutation, which abrogates Yku-TLC1 interaction, resulted in ~500 bp shortening of over-elongated *rif1Δ* and *rif2Δ* telomeres, which is a far bigger effect than in the wild-type background (only 70 bp) [70]. This observation suggests that Rif proteins inhibit Yku-mediated telomere lengthening. Altogether, these data support the idea for the role of Rif proteins in destabilizing Yku-Sir4 binding to telomeres, and hence, transient telomerase recruitment in G1-phase. It must be noted that the *tlc1Δ48* mutation does not completely suppress the telomere over-elongation phenotype of *rif1Δ* and *rif2Δ* mutants, implying that the negative role of Rif proteins in telomere length maintenance is also linked to the Cdc13-Est1 mechanism of telomerase recruitment.

Therefore, it is possible that G1-specific inhibition of telomerase function is controlled by Rif1 and Rif2 proteins at two different levels: (1) by ensuring that G-tails are not produced in G1, and hence, the Cdc13-Est1 telomerase recruitment pathway is not available; and (2) by out-competing Sir4-Yku from telomeres, therefore destabilizing telomerase association with telomeres.

As we can see, Sir4-Yku interaction does not provide stable telomerase recruitment in G1 and is mostly dispensable for telomere maintenance. Then why having a specialized but unproductive telomerase recruitment pathway in G1? We suggest that via short-lived interactions, cells could ensure that telomerase does not localize to sites of DSB repair and engage in unsanctioned repair mechanisms. Indeed, sites of telomere elongation and DSB repair were shown to be spatially separated between different nuclear compartments [113, 114]. Although in this scenario, telomerase is kept at telomeres in G1, it does not engage in productive interactions with telomeres, possibly due to the competition between Rif and Sir proteins for Rap1 binding. Interestingly, the fact that Sir4-Yku80 complexes locate distally from telomere ends [73], might also ensure that telomerase recruitment to telomeres in G1 remains unproductive.

7. Inhibition of telomerase function at DSBs

Due to a similar structure of telomeres and DSBs, telomerase has a potential to add telomeres to broken DNA, a phenomenon called *de novo* telomere addition. The consequences of telomere addition to DSBs could be fatal, because the chromosome fragment distal to the DSB is either lost or can initiate further genomic rearrangements [115, 116]. On the other hand, unsanctioned access of DNA repair mechanisms to telomeres will lead to chromosome fusions and massive genome rearrangements with devastating consequences for cell function [117–120].

To ensure that these accidents do not happen, telomeres and DNA DSBs are spatially separated in the nucleus. DSBs are mainly localized to the nuclear pores, whereas telomeres anchor at the nuclear envelope, although exceptions from this general rule can be observed [113, 114]. In addition, local mechanisms operate both at DSBs and telomeres to ensure that telomerase and DSB repair machinery do not mix up their substrates.

7.1 Telomere anchoring at the nuclear envelope

During G1- and S-phase, telomeres are clustered in approximately 8 foci, which are tethered to the nuclear envelope [113, 121]. Telomere tethering requires interaction between telomeric and nuclear envelope proteins. One telomere tethering pathway involves Sir4 interaction with Esc1 and Mps3 nuclear envelope proteins, whereas the other one depends on the Yku70/80 and Mps3 [122–124]. The two telomere tethering pathways are redundant to a certain extent, although the details of their exact functional interaction remain elusive [121].

Unlike Sir4, Yku70 cannot directly bind Mps3 *in vitro* [123]. This indicates that the Yku-mediated telomere anchoring pathway requires a "bridge" between Yku on the telomere side and Mps3 on the nuclear envelope side. It seems that telomerase might function as such a "bridge," at least during S-phase. It was shown, that on the nuclear envelope side, Est1 interacts with Mps3 [125]. In turn, on the telomere side of this "sandwich," TLC1 and Est1 interact with Yku [126]. In G1-phase, these interactions do not contribute to telomere anchoring, suggesting that another factor may mediate Yku-Mps3 bridging and telomere tethering.

An elegant single-telomere extension experiment revealed that telomeres detach from the nuclear envelope during extension [127]. In this assay, a telomerase-deficient recipient mutant with short telomeres was mated with a wild-type donor cell. In the zygote, short recipient telomeres can be efficiently elongated by the donor telomerase. To track short recipient telomeres, Tel5R was tagged with *lacO*/LacI-GFP, and the nuclear envelope was marked with Nup49-GFP. It was observed that the short Tel5R telomere was released from the nuclear periphery in the first S-phase after mating, when telomere elongation was expected to happen. However, in the second cell cycle, when the short telomere should already have been extended, Tel5R stayed anchored to the nuclear envelope. Notably, in the telomerase-negative zygote, the short telomere did not detach from the nuclear envelope in the first cell cycle. This indicates that telomerase function or the process of telomere elongation in general somehow triggers telomere dissociation from the nuclear envelope.

Deletion of *SIZ2*, encoding for the SUMO E3 ligase, causes long untethered telomeres. This indicates that telomere anchoring to the nuclear envelope is achieved via Siz2-dependent sumoylation of certain tethering components. Accordingly, Siz2 sumoylates Yku70/80 and Sir4 *in vivo*. Moreover, fusing Yku70 with SUMO suppresses the *siz2*Δ telomere tethering defect.

Altogether, these observations suggest that telomere anchoring may not only separate telomere-extension and DSB repair activities in space, but also act as an additional regulatory mechanism, which ensures cell-cycle restriction of telomere elongation.

Although normally clustered at the nuclear envelope, some telomeres can occasionally localize to nuclear pores, the sites of DSB repair. For instance, when a DSB occurs in a subtelomeric region, it must re-localize to the nuclear pore for efficient DSB repair [128]. In addition, eroded telomeres in telomerase-negative cells were also shown to re-localize to nuclear pores [129]. As a result, such critically short telomeres can undergo recombination-dependent elongation that allows cell survival in the absence of telomerase [130].

7.2 Local mechanisms restricting telomerase access to DSBs

De novo telomere addition at DSBs occurs mostly at telomere-like TG-rich regions [131–134]. However, the molecular mechanism of this process is not very well understood. In particular, it is still unknown how telomerase is recruited to the break. Cdc13 is one of the most likely candidates for this role. Consistently, Cdc13 and telomerase can be detected at both TG-flanked and non-TG DSB ends [135]. The single-stranded DNA binding protein complex RPA is another candidate for recruiting telomerase to DSBs. RPA is abundantly present at processed DSB ends and was also shown to mediate telomerase association with telomeres [136, 137].

Chromosome healing via *de novo* telomere addition is an extremely rare event, suggesting that telomerase function is efficiently inhibited at DSBs [115, 134]. In yeast, two mechanisms are implicated in the inhibition of *de novo* telomere addition. The first one involves Mec1-mediated Cdc13 phosphorylation at the S306 residue [138]. In cells expressing the Cdc13-S306A unphosphorylatable protein, the frequency of *de novo* telomere addition to DSBs is increased. This Cdc13 modification does not prevent Cdc13-Est1 interaction, but rather affects Cdc13 association with the breaks. Accordingly, *cdc13-S306A* mutants have increased accumulation of Cdc13 at DSBs. The Mec1-dependent phosphorylation of Cdc13-S306 can be reversed by Pph3, the yeast protein phosphatase 4, and Rrd1, the yeast ortholog of human phosphotyrosyl phosphatase activator. Reversible Cdc13 phosphorylation kept under the control of the DNA damage checkpoint kinase is reminiscent of Cdc13 regulation at telomeres (discussed in Section 4). Therefore, by targeting different phosphorylation sites, cells modulate Cdc13 functions both at telomeres and DSBs.

Another mechanism inhibiting telomerase at DSBs relies on Pif1, a multifunctional 5′-3′ helicase, expressed both in the nucleus and mitochondria [139]. The nuclear Pif1 isoform contributes to DNA replication and repair by promoting Okazaki fragment processing, resolution of G-quadruplex DNA structures, ribosomal DNA replication and DSB repair via Rad51-dependent break-induced replication. In addition, Pif1 inhibits telomere addition both at telomeres and DSBs. Accordingly, cells expressing the *pif1-m2* allele that lack the nuclear Pif1 isoform, have long telomeres, whereas overexpression of *PIF1* leads to telomere shortening [134]. The frequency of *de novo* telomere addition to spontaneous and HO-induced DSBs is increased 200-1000-fold in the absence of Pif1 [131, 132, 134]. The inhibitory effect of Pif1 on *de novo* telomere addition is telomerase-dependent, as inactivation of telomerase function suppresses the high rate of gross chromosomal rearrangements and long telomere phenotype observed in *pif1-m2* cells [132, 140].

Similar to Cdc13, Pif1 function can be modulated by specific phosphorylation events. It was shown that Mec1-dependent phosphorylation specifically regulates Pif1 activity at DSBs [141]. Cells expressing Pif1-4A, which cannot be phosphorylated by Mec1, have the same level of *de novo* telomere addition as *pif1-m2* cells. However, additional telomere lengthening associated with *pif1-m2* does not occur in *pif1-4A* cells. Hence, Mec1-dependent Pif1 phosphorylation is required for telomerase inhibition specifically at DSBs.

In vitro, Pif1 preferentially unwinds RNA-DNA structures [142]. In addition, the helicase activity of Pif1 is required for its telomeric function [140]. Therefore, Pif1 role in inhibition of telomere elongation could be explained by its ability to disrupt the base-pairing between the TLC1 RNA and telomeric DNA, leading to displacement of telomerase from telomeres. Consistent with this model, Pif1 expression is cell-cycle regulated and reaches its maximum during late S-phase, coinciding with the time of telomerase action at telomeres [51, 143]. In addition, Pif1 overexpression reduces telomerase association with telomeres [144]. However, the novel function of the TERRA RNA in recruitment of telomerase clusters to telomeres raises the possibility that TERRA might also be a target for the Pif1 helicase activity, which would explain its negative effect on telomere lengthening [109].

Replication forks can frequently stall at TG-rich sequences, resulting in formation of DSBs flanked with short TG-sequences. At the same time, native telomeres can also reach a critically short length. Despite this structural resemblance, the first type of substrate must be channeled into proper DSB repair mechanisms, whereas telomerase must be inhibited. In turn, short telomeres must not be recognized by DNA repair mechanisms, and instead, must be elongated by telomerase.

How does a cell distinguish between these very similar structures? Recent evidence suggests that cooperation between Cdc13 and Pif1 might channel particular substrates into proper repair pathways [145]. It was shown, that if either a natural telomere or a TG-seed flanking a DSB is shorter than 35-40 bp, it is recognized as a break and is protected from telomerase activity by Pif1. As a result, such a substrate will be processed by DSB repair mechanisms. In turn, when the length of the TG-tract exceeds the 35-40 bp threshold, it is considered as a short telomere, irrespective of whether it is at a break or at the natural chromosome end. Pif1 will not protect such a substrate from telomerase function, thus making it available for telomerase-mediated elongation.

Mec1-dependent Pif1 phosphorylation, which specifically modulates its function at the breaks, does not make Pif1 selectively sensitive to shorter TG-sequences and insensitive to the longer ones [141, 145]. Instead, it was shown that Pif1 selectivity for different TG-substrates is mediated by Cdc13 [145]. If Cdc13 binding to DSBs is attenuated, longer telomeres become sensitive to Pif1. This implies that Cdc13 protects longer TG-sequences from the inhibitory Pif1 activity, thus allowing telomere addition. Interestingly, this effect of Cdc13 is not linked to its role in telomerase recruitment. The Cdc13-Est1 or Cdc13-Est2 fusions did not result in more frequent telomere addition at short TG-sequences when Pif1 was present. It suggests that the role of Cdc13 at the threshold DSB TG-sequences is not due to telomerase recruitment, but could possibly be explained by other Cdc13 functions, such as telomere capping or promoting the C-strand synthesis.

Pif1 was previously implicated in promoting resection in *cdc13-1* cells, where Cdc13 is inactive [146, 147]. Hence, Strecker et al. suggested that at short TG-stretches, less Cdc13 may allow more extensive processing by Pif1 and telomerase inhibition [145]. Once telomere length is over the threshold, it can be protected from Pif1 resection and thus Pif1 no longer inhibits telomerase.

We suggest an alternative explanation that may explain the Pif1 and Cdc13 effect on telomere addition to threshold substrates by their role in Okazaki fragment synthesis. When the replicative helicase runs off telomeric DNA ends ahead of the replisome, single-stranded DNA stretches may become exposed. The CA-rich strand will be rapidly converted into double-stranded DNA by leading-strand synthesis, whereas the TG-rich strand must be filled-in by the lagging-strand replication. It is possible, that if the TG-tract is shorter than 35–40 bp, lagging-strand initiation will be inefficient. In this scenario, the TG-tract, as well as the upstream non-TG sequence, will remain single-stranded. RPA bound to the non-TG sequence will signal for the DNA damage response activation, resulting in the engagement of the 3′-overhang in homology search and DSB repair. However, when the TG-sequence is longer than the threshold, priming for lagging-strand synthesis may be efficient. As a result, single-stranded DNA will get converted into a double-stranded DNA tract, and RPA displaced from the non-TG sequence will no longer signal for the DNA damage response activation. Finally, Cdc13 bound to the leftover single-stranded TG-sequence will recruit telomerase.

Why would the initiation of the lagging-strand synthesis be inefficient at short TG-tracts? In mammals, the CST complex cannot bind single-stranded DNA tracts that are shorter than 32 bp [148]. This suggests that at shorter TG-tracts, the CST complex might not be stable, and hence, cannot efficiently recruit/stabilize DNA polymerase α/primase complex. Consequently, Okazaki fragment synthesis will be inefficient, and the RNA primer could be displaced from DNA via the Pif1 helicase activity. Therefore, the RPA-bound 3′-overhang containing non-TG and TG-sequences will persist and signal for DNA damage response activation. Long TG-tracts, on the other hand, will efficiently accommodate the CST complex, resulting in stable recruitment of the lagging-strand synthesis machinery. The RNA primer would be quickly extended into a full Okazaki fragment, making it inaccessible for Pif1. This can explain, why at longer TG-tracts, telomere addition is not under Pif1 control. According to this scenario, abolishing lagging-strand DNA synthesis, but not RNA priming, should render long TG-tracts sensitive to Pif1, leading to inhibition of telomere addition.

Therefore, at telomeres and DSBs, Pif1 might act at two consecutive steps. First, Pif1 may decide the fate of the telomeric end, and either channel it to the DSB repair pathway or leave it accessible for telomere elongation. As discussed above, this could be due to the Pif1 role in Okazaki fragment synthesis, rather than direct telomerase displacement. At the next step, Pif1 might regulate the extent of telomere elongation by telomerase. Accordingly, Pif1 was shown to inhibit telomerase preferentially at long telomeres [149]. Unlike its function at the threshold TG-tracts, this effect could be due to Pif1 stripping telomerase from telomeric 3′-overhangs.

Although most DSBs are localized to nuclear pores to be repaired, some persistent breaks are brought to the same regions as telomeres [114]. If such a break is deemed "nontelomeric," it will subsequently re-localize to the nuclear pore. Why are these "questionable" DSBs brought so dangerously close to the telomeric sites?

Such behavior can be nicely explained by the role of Pif1 in sorting DSBs from telomeres based on their threshold TG-sequence [145]. By extension, this Pif1-mediated process should co-localize with telomeric clusters tethered to the nuclear envelope.

8. Conclusions and future perspectives

In this review, we have outlined the main regulatory mechanisms that tightly cooperate in order to control telomerase function at telomere and at the breaks. Although our knowledge on this subject is quite extensive, many questions remain. Do the Pop1, Pop6 and Pop7 proteins functionally interact with any other telomerase subunits or telomeric proteins, in addition to Est1 and Est2 (e.g., Est3, Cdc13, Yku). How do the Pop proteins affect telomerase composition and function during cell cycle? Are there any unknown telomerase subunits that are yet to be discovered, and if so, what are their roles? What is the functional significance of the Sir4-Yku-TLC1 telomerase recruitment pathway? Is it related to the sequestration of telomerase from the DSB repair sites, as we proposed above, or does it play another role? Telomere anchoring at the nuclear periphery seems to affect their elongation. How is this process mediated, and what is the significance of telomerase in telomere anchoring at the nuclear envelope? What is the mechanism that allows Pif1 to distinguish DSBs from short telomeres and inhibit telomerase only at DSBs? All these questions and many more wait for answers, which will help to understand better the intricate network of telomerase regulatory pathways.

Acknowledgements

We thank Emeline Pasquier and Erin Bonnell for input on this manuscript. Research in our lab is supported by the grant from the Canadian Institutes of Health Research (FDN 154315), the Canadian Research Chair in Telomere Biology (to R.J.W.), the postdoctoral fellowship of Fonds de la recherche en santé du Québec (to Y.V.) and the Abdenour-Nabid scholarship from the Faculty of Medicine and Health Sciences, Université de Sherbrooke (to A.K.).

Conflict of interest

The authors declare no conflict of interest.

Author details

Yulia Vasianovich, Alexandra Krallis and Raymund Wellinger*
Université de Sherbrooke, Sherbrooke, Canada

*Address all correspondence to: raymund.wellinger@usherbrooke.ca

References

[1] Olovnikov A. Principle of marginotomy in the synthesis of polynucleotides at a template. Doklady. Biochemistry and Biophysics. 1971;**201**:394-397

[2] Watson JD. Origin of concatemeric T7 DNA. Nature: New Biology. 1972;**239**:197

[3] Wellinger RJ, Ethier K, Labrecque P, et al. Evidence for a new step in telomere maintenance. Cell. 1996;**85**:423-433

[4] Soudet J, Jolivet P, Teixeira MT. Elucidation of the DNA end-replication problem in *Saccharomyces cerevisiae*. Molecular Cell. 2014;**53**:954-964

[5] Wellinger RJ. In the end, what's the problem? Molecular Cell. 2014;**53**:855-856

[6] Hayflick L. The limited *in vitro* lifetime of human diploid cell strains. Experimental Cell Research. 1965;**37**:614-636

[7] Harley C, Futcher A, Greider C. Telomeres shorten during ageing of human fibroblasts. Nature. 1990;**345**:458-460

[8] Wellinger RJ, Zakian VA. Everything you ever wanted to know about *Saccharomyces cerevisiae* telomeres: Beginning to end. Genetics. 2012;**191**:1073-1105

[9] Greider CW, Blackburn EH. Identification of a specific telomere terminal transferase activity in *Tetrahymena* extracts. Cell. 1985;**43**:405-413

[10] Greider CW, Blackburn EH. The telomere terminal transferase of *Tetrahymena* is a ribonucleoprotein enzyme with two kinds of primer specificity. Cell. 1987;**51**:887-898

[11] Savage SA, Bertuch AA. The genetics and clinical manifestations of telomere biology disorders. Genetics in Medicine. 2010;**12**:753-764

[12] Lingner J, Hughes TR, Shevchenko A, et al. Reverse transcriptase motifs in the catalytic subunit of telomerase. Science. 1997;**276**:561-567

[13] Lingner J, Cech TR, Hughes TR, et al. Three ever shorter telomere (EST) genes are dispensable for *in vitro* yeast telomerase activity. Proceedings of the National Academy of Sciences. 1997;**94**:11190-11195

[14] Lingner J, Cech TR. Purification of telomerase from *Euplotes aediculatus*: Requirement of a primer 3′ overhang. Proceedings of the National Academy of Sciences of the United States of America. 1996;**93**:10712-10717

[15] Harrington L, Zhou W, McPhail T, et al. Human telomerase contains evolutionarily conserved catalytic and structural subunits. Genes & Development. 1997;**11**:3109-3115

[16] Kilian A, Bowtell DDL, Abud HE, et al. Isolation of a candidate human telomerase catalytic subunit gene, which reveals complex splicing patterns in different cell types. Human Molecular Genetics. 1997;**6**:2011-2019

[17] Meyerson M, Counter CM, Ng Eaton E, et al. hEST2, the putative human telomerase catalytic subunit gene, is up-regulated in tumor cells and during immortalization. Cell. 1997;**90**:785-795

[18] Nakamura TM, Morin GB, Chapman KB, et al. Telomerase catalytic subunit homologs from fission yeast and human. Science. 1997;**277**:955-959

[19] Bryan TM, Sperger JM, Chapman KB, et al. Telomerase reverse

transcriptase genes identified in *Tetrahymena thermophila* and Oxytricha trifallax. Biochemistry. 1998;**95**:8479-8484

[20] Collins K, Gandhi L. The reverse transcriptase component of the *Tetrahymena* telomerase ribonucleoprotein complex. Proceedings of the National Academy of Sciences. 1998;**95**:8485-8490

[21] Greenberg RA, Allsopp RC, Chin L, et al. Expression of mouse telomerase reverse transcriptase during development, differentiation and proliferation. Oncogene. 1998;**16**:1723-1730

[22] Evans SK, Lundblad V. Est1 and Cdc13 as comediators of telomerase access. Science. 1999;**286**:117-120

[23] Hughes TR, Evans SK, Weilbaecher RG, et al. The Est3 protein is a subunit of yeast telomerase. Current Biology. 2000;**10**:809-812

[24] Lee J, Mandell EK, Rao T, et al. Investigating the role of the Est3 protein in yeast telomere replication. Nucleic Acids Research. 2010;**38**:2279-2290

[25] Lemieux B, Laterreur N, Perederina A, et al. Active yeast telomerase shares subunits with ribonucleoproteins RNase P and RNase MRP. Cell. 2016;**165**:1171-1181

[26] Nugent CI, Hughes TR, Lue NF, et al. Cdc13p: A single-strand telomeric DNA-binding protein with a dual role in yeast telomere maintenance. Science. 1996;**274**:249-252

[27] Pennock E, Buckley K, Lundblad V. Cdc13 delivers separate complexes to the telomere for end protection and replication. Cell. 2001;**104**:387-396

[28] Peterson SE, Stellwagen AE, Diede SJ, et al. The function of a stem-loop in telomerase RNA is linked to the DNA repair protein Ku. Nature Genetics. 2001;**27**:64-67

[29] Seto AG, Zaug AJ, Sobel SG, et al. *Saccharomyces cerevisiae* telomerase is an Sm small nuclear ribonucleoprotein particle. Nature. 1999;**401**:177-180

[30] Fisher TS, Taggart AKP, Zakian VA. Cell cycle-dependent regulation of yeast telomerase by Ku. Nature Structural & Molecular Biology. 2004;**11**:1198-1205

[31] Gallardo F, Olivier C, Dandjinou AT, et al. TLC1 RNA nucleo-cytoplasmic trafficking links telomerase biogenesis to its recruitment to telomeres. The EMBO Journal. 2008;**27**:748-757

[32] Dandjinou AT, Lévesque N, Larose S, et al. A phylogenetically based secondary structure for the yeast telomerase RNA. Current Biology. 2004;**14**:1148-1158

[33] Livengood AJ, Zaug AJ, Cech TR. Essential regions of *Saccharomyces cerevisiae* telomerase RNA: Separate elements for Est1p and Est2p interaction. Molecular and Cellular Biology. 2002;**22**:2366-2374

[34] Chappell AS, Lundblad V. Structural elements required for association of the *Saccharomyces cerevisiae* telomerase RNA with the Est2 reverse transcriptase. Molecular and Cellular Biology. 2004;**24**:7720-7736

[35] Seto AG, Livengood AJ, Tzfati Y, et al. A bulged stem tethers Est1p to telomerase RNA in budding yeast. Genes & Development. 2002;**16**:2800-2812

[36] Lubin JW, Tucey TM, Lundblad V. The interaction between the yeast telomerase RNA and the Est1 protein requires three structural elements. RNA. 2012;**18**:1597-1604

[37] Laterreur N, Lemieux B, Neumann H, et al. The yeast telomerase module

for telomere recruitment requires a specific RNA architecture. RNA. 2018;**24**:1067-1079

[38] Evans SK, Lundblad V. The Est1 subunit of *Saccharomyces cerevisiae* telomerase makes multiple contributions to telomere length maintenance. Genetics. 2002;**162**:1101-1115

[39] Tucey TM, Lundblad V. Regulated assembly and disassembly of the yeast telomerase quaternary complex. Genes & Development. 2014;**28**:2077-2089

[40] Lubin JW, Tucey TM, Lundblad V. Using separation-of-function mutagenesis to define the full spectrum of activities performed by the Est1 telomerase subunit *in vivo*. Genetics. 2018;**208**:97-110

[41] Friedman KL, Heit JJ, Long DM, et al. N-terminal domain of yeast telomerase reverse transcriptase: Recruitment of Est3p to the telomerase complex. Molecular Biology of the Cell. 2003;**14**:1-13

[42] Talley JM, DeZwaan DC, Maness LD, et al. Stimulation of yeast telomerase activity by the ever shorter telomere 3 (Est3) subunit is dependent on direct interaction with the catalytic protein Est2. The Journal of Biological Chemistry. 2011;**286**:26431-26439

[43] Yen W-F, Chico L, Lei M, et al. Telomerase regulatory subunit Est3 in two *Candida* species physically interacts with the TEN domain of TERT and telomeric DNA. Proceedings of the National Academy of Sciences. 2011;**108**:20370-20375

[44] Jacobs SA, Podell ER, Cech TR. Crystal structure of the essential N-terminal domain of telomerase reverse transcriptase. Nature Structural & Molecular Biology. 2006;**13**:218-225

[45] Diede SJ, Gottschling DE. Telomerase-mediated telomere addition *in vivo* requires DNA primase and DNA polymerases alpha and delta. Cell. 1999;**99**:723-733

[46] Marcand S, Brevet V, Mann C, et al. Cell cycle restriction of telomere elongation. Current Biology. 2000;**10**:487-490

[47] Rao T, Lubin JW, Armstrong GS, et al. Structure of Est3 reveals a bimodal surface with differential roles in telomere replication. Proceedings of the National Academy of Sciences. 2014;**111**:214-218

[48] Osterhage JL, Talley JM, Friedman KL. Proteasome-dependent degradation of Est1p regulates the cell cycle-restricted assembly of telomerase in Saccharomyces cerevisiae. Nature Structural & Molecular Biology. 2006;**13**:720-728

[49] Esakova O, Krasilnikov AS. Of proteins and RNA: The RNase P/MRP family. RNA. 2010;**16**:1725-1747

[50] Dionne I, Wellinger RJ. Processing of telomeric DNA ends requires the passage of a replication fork. Nucleic Acids Research. 1998;**26**:5365-5371

[51] Taggart AKP, Teng S-C, Zakian VA. Est1p as a cell cycle-regulated activator of telomere-bound telomerase. Science. 2002;**297**:1023-1026

[52] Gallardo F, Laterreur N, Cusanelli E, et al. Live cell imaging of telomerase RNA dynamics reveals cell cycle-dependent clustering of telomerase at elongating telomeres. Molecular Cell. 2011;**44**:819-827

[53] Ferguson JL, Chao WCH, Lee E, et al. The anaphase promoting complex contributes to the degradation of the *S. cerevisiae* telomerase recruitment subunit Est1p. PLoS One. 2013;**8**(1):e55055

[54] Tucey TM, Lundblad V. A yeast telomerase complex containing the

Est1 recruitment protein is assembled early in the cell cycle. Biochemistry. 2013;**52**:1131-1133

[55] Spellman PT, Sherlock G, Zhang MQ, et al. Comprehensive identification of cell cycle-regulated genes of the yeast *Saccharomyces cerevisiae* by microarray hybridization. Molecular and Cellular Biology. 1998;**9**:3273-3297

[56] Larose S, Laterreur N, Ghazal G, et al. RNase III-dependent regulation of yeast telomerase. The Journal of Biological Chemistry. 2007;**282**:4373-4381

[57] Lin KW, McDonald KR, Guise AJ, et al. Proteomics of yeast telomerase identified Cdc48-Npl4-Ufd1 and Ufd4 as regulators of Est1 and telomere length. Nature Communications. 2015;**6**:1-14

[58] Finley D, Ulrich HD, Sommer T, et al. The ubiquitin-proteasome system of *Saccharomyces cerevisiae*. Genetics. 2012;**192**:319-360

[59] Fell VL, Schild-Poulter C. The Ku heterodimer: Function in DNA repair and beyond. Mutation Research. 2015;**763**:15-29

[60] Walker JR, Corpina RA, Goldberg J. Structure of the Ku heterodimer bound to DNA and its implications for double-strand break repair. Nature. 2001;**412**:607-614

[61] Gravel S, Larrivée M, Labrecque P, et al. Yeast Ku as a regulator of chromosomal DNA end structure. Science. 1998;**280**:741-744

[62] Polotnianka RM, Li J, Lustig AJ. The yeast Ku heterodimer is essential for protection of the telomere against nucleolytic and recombinational activities. Current Biology. 1998;**8**:831-835

[63] Boulton SJ, Jackson SP. Components of the Ku-dependent non-homologous end-joining pathway are involved in telomeric length maintenance and telomeric silencing. The EMBO Journal. 1998;**17**:1819-1828

[64] Mishra K, Shore D. Yeast Ku protein plays a direct role in telomeric silencing and counteracts inhibition by Rif proteins. Current Biology. 1999;**9**:1123-1126

[65] Laroche T, Martin SG, Gotta M, et al. Mutation of yeast Ku genes disrupts the subnuclear organization of telomeres. Current Biology. 1998;**8**:653-657

[66] Stellwagen AE, Haimberger ZW, Veatch JR, et al. Ku interacts with telomerase RNA to promote telomere addition at native and broken chromosome ends. Genes & Development. 2003;**17**:2384-2395

[67] Chen H, Xue J, Churikov D, et al. Structural insights into yeast telomerase recruitment to telomeres. Cell. 2018;**172**:331-343

[68] Pfingsten JS, Goodrich KJ, Taabazuing C, et al. Mutually exclusive binding of telomerase RNA and DNA by Ku alters telomerase recruitment model. Cell. 2012;**148**:922-932

[69] Roy R, Meier B, McAinsh AD, et al. Separation-of-function mutants of yeast Ku80 reveal a Yku80p-Sir4p interaction involved in telomeric silencing. The Journal of Biological Chemistry. 2004;**279**:86-94

[70] Hass EP, Zappulla DC. The Ku subunit of telomerase binds Sir4 to recruit telomerase to lengthen telomeres in *S. cerevisiae*. eLife. 2015;**4**:1-19

[71] Lopez CR, Ribes-Zamora A, Indiviglio SM, et al. Ku must load directly onto the chromosome end in order to mediate its telomeric functions. PLoS Genetics. 2011;7(8):e1002233

[72] Ribes-Zamora A, Mihalek I, Lichtarge O, et al. Distinct faces of the Ku heterodimer mediate DNA repair and telomeric functions. Nature Structural & Molecular Biology. 2007;**14**:301-307

[73] Larcher MV, Pasquier E, MacDonald RS, et al. Ku binding on telomeres occurs at sites distal from the physical chromosome ends. PLoS Genetics. 2016;**12**:1-29

[74] Wellinger RJ, Wolf AJ, Zakian VA. *Saccharomyces* telomeres acquire single-strand TG1-3 tails late in S phase. Cell. 1993;**72**:51-60

[75] Wellinger RJ, Wolf AJ, Zakian VA. Origin activation and formation of single-strand TG1-3 tails occur sequentially in late S phase on a yeast linear plasmid. Molecular and Cellular Biology. 1993;**13**:4057-4065

[76] Larrivée M, LeBel C, Wellinger RJ. The generation of proper constitutive G-tails on yeast telomeres is dependent on the MRX complex. Genes & Development. 2004;**18**:1391-1396

[77] Bonetti D, Martina M, Clerici M, et al. Multiple pathways regulate 3′ overhang generation at *S. cerevisiae* telomeres. Molecular Cell. 2009;**35**:70-81

[78] Faure V, Coulon S, Hardy J, et al. Cdc13 and telomerase bind through different mechanisms at the lagging- and leading-strand telomeres. Molecular Cell. 2010;**38**:842-852

[79] Qi H, Zakian VA. The *Saccharomyces* telomere-binding protein Cdc13p interacts with both the catalytic subunit of DNA polymerase alpha and the telomerase-associated Est1 protein. Genes & Development. 2000;**14**:1777-1788

[80] Wu Y, Zakian VA. The telomeric Cdc13 protein interacts directly with the telomerase subunit Est1 to bring it to telomeric DNA ends *in vitro*. Proceedings of the National Academy of Sciences. 2011;**108**:20362-20369

[81] Chan A, Boulé J-B, Zakian VA. Two pathways recruit telomerase to *Saccharomyces cerevisiae* telomeres. PLoS Genetics. 2008;**4**:1-11

[82] Tseng SF, Lin JJ, Teng SC. The telomerase-recruitment domain of the telomere binding protein Cdc13 is regulated by Mec1p/Tel1p-dependent phosphorylation. Nucleic Acids Research. 2006;**34**:6327-6336

[83] Shen Z-J, Hsu P-H, Su Y-T, et al. PP2A and Aurora differentially modify Cdc13 to promote telomerase release from telomeres at G2/M phase. Nature Communications. 2014;**5**:5312

[84] Gao H, Toro TB, Paschini M, et al. Telomerase recruitment in *Saccharomyces cerevisiae* is not dependent on Tel1-mediated phosphorylation of Cdc13. Genetics. 2010;**186**:1147-1159

[85] Wu Y, DiMaggio PA, Perlman DH, et al. Novel phosphorylation sites in the *S. cerevisiae* Cdc13 protein reveal new targets for telomere length regulation. Journal of Proteome Research. 2013;**12**:316-327

[86] Puglisi A, Bianchi A, Lemmens L, et al. Distinct roles for yeast Stn1 in telomere capping and telomerase inhibition. The EMBO Journal. 2008;**27**:2328-2339

[87] Chandra A, Hughes TR, Nugent CI, et al. Cdc13 both positively and negatively regulates telomere replication. Genes & Development. 2001;**15**:404-414

[88] Grossi S, Puglisi A, Dmitriev PV, et al. Pol12, the B subunit of DNA polymerase α, functions in both telomere capping and length

regulation. Genes & Development. 2004;**18**:992-1006

[89] Liu C-C, Gopalakrishnan V, Poon L-F, et al. Cdk1 regulates the temporal recruitment of telomerase and Cdc13-Stn1-Ten1 complex for telomere replication. Molecular and Cellular Biology. 2014;**34**:57-70

[90] Gopalakrishnan V, Tan CR, Li S. Sequential phosphorylation of CST subunits by different cyclin-Cdk1 complexes orchestrate telomere replication. Cell Cycle. 2017;**16**:1271-1287

[91] Teixeira MT, Arneric M, Sperisen P, et al. Telomere length homeostasis is achieved via a switch between telomerase-extendible and -nonextendible states. Cell. 2004;**117**:323-335

[92] Marcand S. A protein-counting mechanism for telomere length regulation in yeast. Science. 1997;**275**:986-990

[93] Levy DL, Blackburn EH. Counting of Rif1p and Rif2p on *Saccharomyces cerevisiae* telomeres regulates telomere length. Molecular and Cellular Biology. 2004;**24**:10857-10867

[94] Hardy CFJ, Sussel L, Shore D. A RAP1-interacting protein involved in transcriptional silencing and telomere length regulation. Genes & Development. 1992;**6**:801-814

[95] Wotton D, Shore D. A novel Rap1p-interacting factor, Rif2p, cooperates with Rif1p to regulate telomere length in Saccharomyces cerevisiae. Genes & Development. 1997;**11**:748-760

[96] Hirano Y, Fukunaga K, Sugimoto K. Rif1 and Rif2 inhibit localization of Tel1 to DNA ends. Molecular Cell. 2009;**33**:312-322

[97] Bianchi A, Shore D. Increased association of telomerase with short telomeres in yeast. Genes & Development. 2007;**21**:1726-1730

[98] Hector RE, Shtofman RL, Ray A, et al. Tel1p preferentially associates with short telomeres to stimulate their elongation. Molecular Cell. 2007;**27**:851-858

[99] Sabourin M, Tuzon CT, Zakian VA. Telomerase and Tel1p preferentially associate with short telomeres in *S. cerevisiae*. Molecular Cell. 2007;**27**:550-561

[100] Goudsouzian LK, Tuzon CT, Zakian VA. *S. cerevisiae* Tel1p and Mre11p are required for normal levels of Est1p and Est2p telomere association. Molecular Cell. 2006;**24**:603-610

[101] Nakada D, Matsumoto K, Sugimoto K. ATM-related Tel1 associates with double-strand breaks through an Xrs2-dependent mechanism. Genes & Development. 2003;**17**:1957-1962

[102] Deng SK, Yin Y, Petes TD, et al. Mre11-Sae2 and RPA collaborate to prevent palindromic gene amplification. Molecular Cell. 2015;**60**:500-508

[103] Martina M, Clerici M, Baldo V, et al. A balance between Tel1 and Rif2 activities regulates nucleolytic processing and elongation at telomeres. Molecular and Cellular Biology. 2012;**32**:1604-1617

[104] Bonetti D, Clerici M, Anbalagan S, et al. Shelterin-like proteins and Yku inhibit nucleolytic processing of *Saccharomyces cerevisiae* telomeres. PLoS Genetics. 2010;**6**:1-14

[105] Bonetti D, Clerici M, Manfrini N, et al. The MRX complex plays multiple functions in resection of Yku- and Rif2-protected DNA ends. PLoS One. 2010;**5**(11):e14142

[106] Greider CW. Regulating telomere length from the inside out: The

replication fork model. Genes & Development. 2016;**30**:1483-1491

[107] Azzalin CM, Reichenbach P, Khoriauli L, et al. Telomeric repeat-containing RNA and RNA surveillance factors at mammalian chromosome ends. Science. 2007;**318**:798-801

[108] Schoeftner S, Blasco MA. Developmentally regulated transcription of mammalian telomeres by DNA-dependent RNA polymerase II. Nature Cell Biology. 2008;**10**:228-236

[109] Cusanelli E, Romero CAP, Chartrand P. Telomeric noncoding RNA TERRA is induced by telomere shortening to nucleate telomerase molecules at short telomeres. Molecular Cell. 2013;**51**:780-791

[110] Graf M, Bonetti D, Lockhart A, et al. Telomere length determines TERRA and R-loop regulation through the cell cycle. Cell. 2017;**170**:72-85.e14

[111] Iglesias N, Redon S, Pfeiffer V, et al. Subtelomeric repetitive elements determine TERRA regulation by Rap1/Rif and Rap1/Sir complexes in yeast. EMBO Reports. 2011;**12**:587-593

[112] Moretti P, Freeman K, Coodly L, et al. Evidence that a complex of SIR proteins interacts with the silencer and telomere-binding protein RAP1. Genes & Development. 1994;**8**:2257-2269

[113] Gotta M, Laroche T, Formenton A, et al. The clustering of telomeres and colocalization with Rap1, Sir3, and Sir4 proteins in wild-type Saccharomyces cerevisiae. The Journal of Cell Biology. 1996;**134**:1349-1363

[114] Gartenberg MR. Life on the edge: Telomeres and persistent DNA breaks converge at the nuclear periphery. Genes & Development. 2009;**13**:1027-1031

[115] Kramer KM, Haber JE. New telomeres in yeast are initiated with a highly selected subset of TG1-3 repeats. Genes & Development. 1993;**7**:2345-2356

[116] Yu GL, Blackburn EH. Developmentally programmed healing of chromosomes by telomerase in *Tetrahymena*. Cell. 1991;**67**:823-832

[117] McClintock B. The stability of broken ends of chromosomes in *Zea mays*. Genetics. 1941;**26**:234-282

[118] Muller HJ. The remaking of chromosomes. The Collecting Net. 1938;**13**:1181-1198

[119] Sandell LL, Zakian VA. Loss of a yeast telomere: Arrest, recovery, and chromosome loss. Cell. 1993;**75**:729-739

[120] Longhese MP. DNA damage response at functional and dysfunctional telomeres. Genes & Development. 2008;**22**:125-140

[121] Hediger F, Neumann FR, Van Houwe G, et al. Live imaging of telomeres. Current Biology. 2002;**12**:2076-2089

[122] Taddei A, Hediger F, Neumaan FR, et al. Separation of silencing from perinuclear anchoring functions in yeast Ku80, Sir4 and Esc1 proteins. The EMBO Journal. 2004;**23**:1301-1312

[123] Bupp JM, Martin AE, Stensrud ES, et al. Telomere anchoring at the nuclear periphery requires the budding yeast Sad1-UNC-84 domain protein Mps3. The Journal of Cell Biology. 2007;**179**:845-854

[124] Andrulis E, Zappulla D. Esc1, a nuclear periphery protein required for Sir4-based plasmid anchoring and partitioning. Molecular and Cellular Biology. 2002;**22**:8292-8301

[125] Antoniacci LM, Kenna MA, Skibbens RV. The nuclear envelope and spindle pole body-associated Mps3

protein bind telomere regulators and function in telomere clustering. Cell Cycle. 2007;**6**:75-79

[126] Schober H, Ferreira H, Kalck V, et al. Yeast telomerase and the SUN domain protein Mps3 anchor telomeres and repress subtelomeric recombination. Genes & Development. 2009;**23**:928-938

[127] Ferreira HC, Luke B, Schober H, et al. The PIAS homologue Siz2 regulates perinuclear telomere position and telomerase activity in budding yeast. Nature Cell Biology. 2011;**13**:867-874

[128] Therizols P, Fairhead C, Cabal GG, et al. Telomere tethering at the nuclear periphery is essential for efficient DNA double strand break repair in subtelomeric region. The Journal of Cell Biology. 2006;**172**:189-199

[129] Khadaroo B, Teixeira MT, Luciano P, et al. The DNA damage response at eroded telomeres and tethering to the nuclear pore complex. Nature Cell Biology. 2009;**11**:980

[130] Churikov D, Charifi F, Eckert-Boulet N, et al. SUMO-dependent relocalization of eroded telomeres to nuclear pore complexes controls telomere recombination. Cell Reports. 2016;**15**:1242-1253

[131] Mangahas JL, Alexander MK, Sandell LL, et al. Repair of chromosome ends after telomere loss in *Saccharomyces*. Molecular Biology of the Cell. 2001;**12**:4078-4089

[132] Myung K, Chen C, Kolodner RD. Multiple pathways cooperate in the suppression of genome instability in *Saccharomyces cerevisiae*. Nature. 2001;**411**:1073-1076

[133] Myung K, Datta A, Kolodner RD. Suppression of spontaneous chromosomal rearrangements by S phase checkpoint functions in *Saccharomyces cerevisiae*. Cell. 2001;**104**:397-408

[134] Schulz VP, Zakian VA. The saccharomyces PIF1 DNA helicase inhibits telomere elongation and *de novo* telomere formation. Cell. 1994;**76**:145-155

[135] Ribaud V, Ribeyre C, Damay P, et al. DNA-end capping by the budding yeast transcription factor and subtelomeric binding protein Tbf1. The EMBO Journal. 2012;**31**:138-149

[136] Luciano P, Coulon S, Faure V, et al. RPA facilitates telomerase activity at chromosome ends in budding and fission yeasts. The EMBO Journal. 2012;**31**:2034-2046

[137] Schramke V, Luciano P, Brevet V, et al. RPA regulates telomerase action by providing Est1p access to chromosome ends. Nature Genetics. 2004;**36**:46-54

[138] Zhang W, Durocher D. *De novo* telomere formation is suppressed by the Mec1-dependent inhibition of Cdc13 accumulation at DNA breaks. Genes & Development. 2010;**24**:502-515

[139] Bochman ML, Sabouri N, Zakian VA. Unwinding the functions of the Pif1 family helicases. DNA Repair (Amst). 2010;**9**:237-249

[140] Zhou J-Q, Monson EK, Teng S-C, et al. Pif1p helicase, a catalytic inhibitor of telomerase in yeast. Science. 2000;**289**:771-774

[141] Makovets S, Blackburn EH. DNA damage signalling prevents deleterious telomere addition at DNA breaks. Nature Cell Biology. 2009;**11**:1383-1386

[142] Boulé JB, Zakian VA. The yeast Pif1p DNA helicase preferentially unwinds RNA-DNA substrates. Nucleic Acids Research. 2007;**35**:5809-5818

[143] Vega LR, Phillips JA, Thornton BR, et al. Sensitivity of yeast strains with long G-tails to levels of telomere-bound telomerase. PLoS Genetics. 2007;**3**:1065-1075

[144] Boulé JB, Vega LR, Zakian VA. The yeast Pif1p helicase removes telomerase from telomeric DNA. Nature. 2005;**438**:57-61

[145] Strecker J, Stinus S, Caballero MP, et al. A sharp Pif1-dependent threshold separates DNA double-strand breaks from critically short telomeres. eLife. 2017;**6**:1-25

[146] Dewar JM, Lydall D. Pif1- and Exo1-dependent nucleases coordinate checkpoint activation following telomere uncapping. The EMBO Journal. 2010;**29**:4020-4034

[147] Garvik B, Carson M, Hartwell L. Single-stranded DNA arising at telomeres in *cdc13* mutants may constitute a specific signal for the RAD9 checkpoint. Molecular and Cellular Biology. 1995;**15**:6128-6138

[148] Miyake Y, Nakamura M, Nabetani A, et al. RPA-like mammalian Ctc1-Stn1-Ten1 complex binds to single-stranded DNA and protects telomeres independently of the Pot1 pathway. Molecular Cell. 2009;**36**:193-206

[149] Phillips JA, Chan A, Paeschke K, et al. The Pif1 helicase, a negative regulator of telomerase, acts preferentially at long telomeres. PLoS Genetics. 2015;**11**:1-19

Chapter 3

Telomere Formation Systems in Budding and Fission Yeasts

Julien Audry and Kurt W. Runge

Abstract

Telomeres are specialized structures essential for genomic stability in eukaryotic cells. Inducible systems causing telomere shortening or telomere formation from short tracts of telomere repeats were developed in the late 1990s in *Saccharomyces cerevisiae* and have been adapted to investigate multiple aspects of telomere biology. In the formation system, an internal tract of telomere repeats is placed next to an inducible double-strand break. Inducing the break converts the telomere tract into a functional telomere whose fate can be followed kinetically and allows one to assay elongation, protein recruitment, and the DNA damage checkpoint activation. This work was extended to *Schizosaccharomyces pombe*, as it shares some features of telomeric chromatin with mammalian cells that are missing in *S. cerevisiae*. The *S. pombe* system has revealed novel aspects of telomeric chromatin formation and similarities with *S. cerevisiae*. This chapter will review these past discoveries in different yeast model organisms, and what they reveal about telomere physiology that may well be conserved in mammals.

Keywords: telomere formation, chromosome end, double-strand break, checkpoint, heterochromatin, yeast, *Saccharomyces cerevisiae*, *Schizosaccharomyces pombe*, I-SceI, HO, endonuclease, Cre, recombinase

1. Introduction

1.1 Context

Telomeres are the physical ends of linear eukaryotic chromosomes. The chromosomal end structure of most telomeres is composed of G-rich DNA repeats bound by specific proteins (**Table 1** and **Figure 1**). One of the major functions of this nucleoprotein complex is to prevent chromosome fusions and the recognition of the ends as a double-strand break (DSB) [1, 2]. Due to the semi-conservative replication and nuclease activities occurring at the chromosome ends, the telomeres shorten at each cell division, a process called the "end replication problem" [3, 4]. To counteract this gradual shortening, telomerase can elongate short telomeres at each division [5, 6]. Telomerase is expressed constitutively in unicellular eukaryotic organisms such as yeasts, but not in most of the somatic cells of humans [7]. The gradual erosion of telomere repeats leads to loss of telomere binding proteins and chromosome end deprotection, which activates the DNA damage response (DDR). The DDR can cause replicative senescence or apoptosis, acting as an anti-proliferative barrier [8, 9]. The dysfunction of this barrier can generate genomic instability, chromosomal aberrations and initiate tumorigenesis or oncogenic transformation of the cells [10].

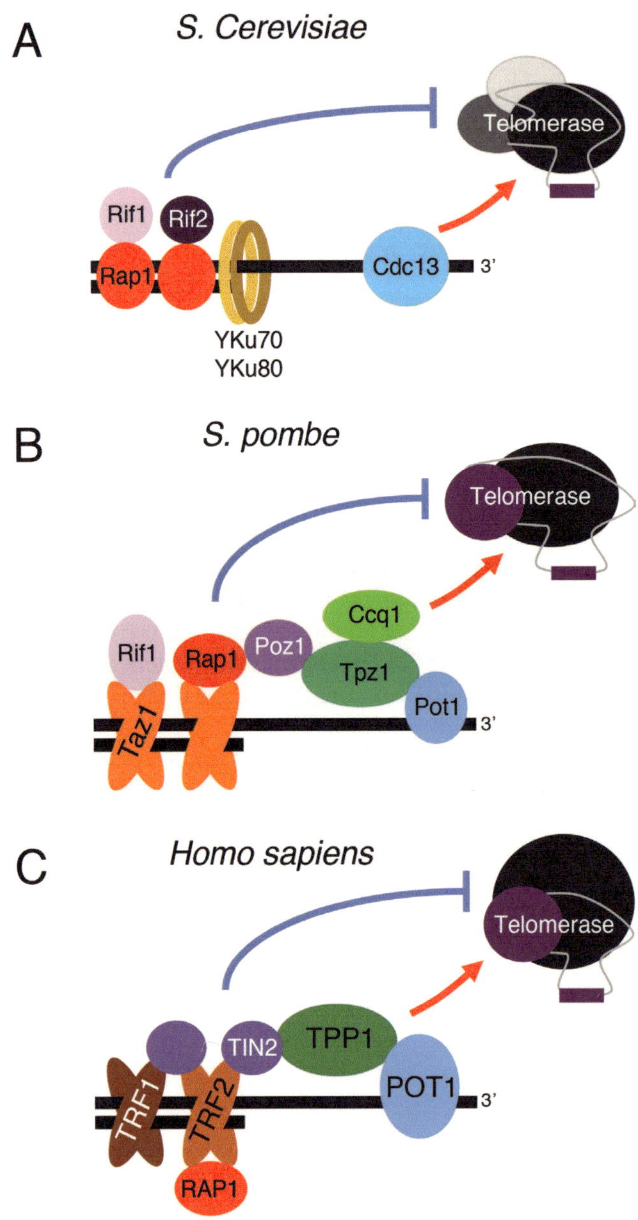

Figure 1.
Telomere complexes conservation from yeasts to humans (A) S. cerevisiae telomere complex. (B) S. pombe telomere complex. (C) Humans telomere complex (also named "shelterin").

Telomere length is defined as the number of short G-rich repeats at chromosome ends. Organisms have a required minimum number of repeats, or telomere repeat tracts, to recruit enough telomere binding proteins to accomplish telomere function [11–15]. Some proteins within the telomere protein complex recruit telomerase to allow elongation of the telomeres (positively regulating telomerase), while others prevent telomerase from continuously adding repeats (negatively regulating telomerase) (**Tables 1** and **2**).

Telomere length is linked to aging, as it was shown that late generation mice with limited telomerase (heterozygotes for telomerase RNA mTR+/−) have short telomeres and revealed a decrease of tissue renewal capacity [16]. In contrast, mice overexpressing telomerase have an extended lifespan [17]. In humans, a correlation

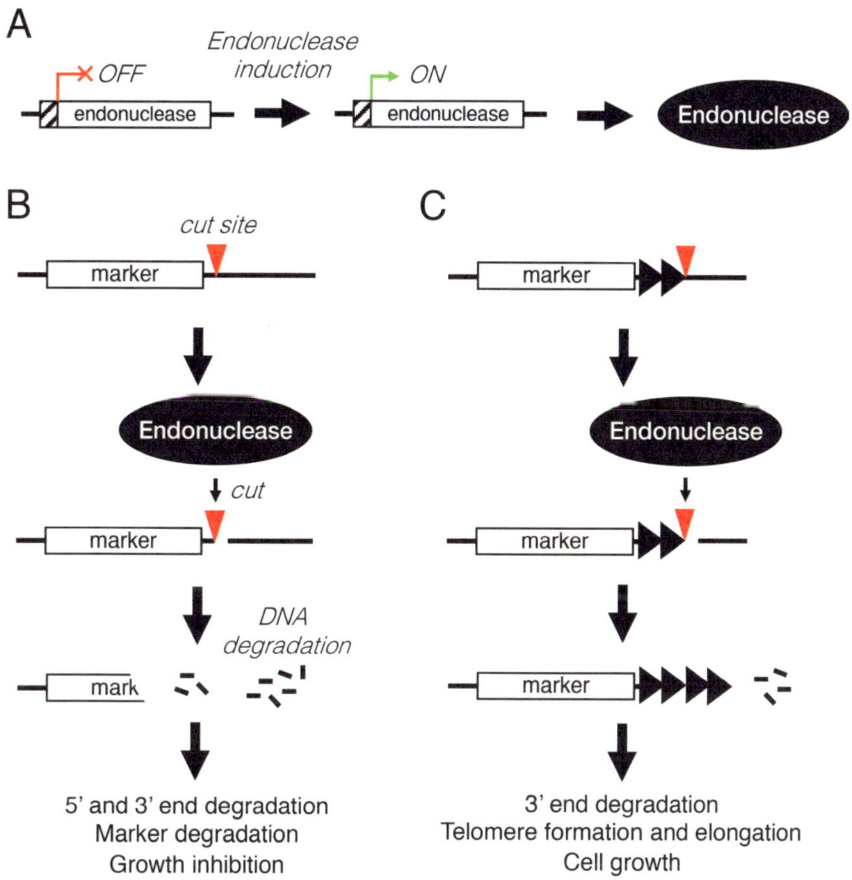

Figure 2.
Double-strand break (DSB) and telomere formation systems (A) restriction enzyme or endonuclease with no natural sites in the genome is produced in cells from a rapidly inducible promoter. (B) Inducible DSB system: a unique cut site (red triangle) is engineered next to a selectable marker gene. After induction, the endonuclease produces a DSB which leads to the degradation of both 5' and 3' strands on each side of the cut site (indicated by short black lines) and to the marker degradation, conducting to cell growth inhibition. (C) Inducible telomere formation system: the cut here exposes telomere repeats (black triangles) to form a new functional telomere that is stable and elongated. If the distal chromosomal DNA (3' to the cut site) is dispensable, the new functional telomere allows normal cell growth. Modified and adapted from Wang et al. [107].

between telomere length, aging, and age-related diseases has also been shown (reviewed in [18, 19]). For instance, in dyskeratosis congenita or its most severe form: Hoyeraal-Hreidarsson syndrome, patients have abnormally short telomeres and shorter lifespan [20]. Most of the known mutations associated are found in telomere proteins (TIN2), telomerase (hTR and hTERT) or the telomerase-associated factor as dyskerin (DKC1), which stabilizes the telomerase RNA (hTR) (**Table 2**) [18]. Moreover, shortened telomere length was observed in other premature aging disorders such as Down's syndrome leading to accelerated aging, or Cockayne syndrome in which neurological degeneration is observed (aging-disorders linked to a telomere phenotype reviewed in [21]). In Down's syndrome, the genetic defect observed is mostly a trisomy of chromosome 21, which increased the biological age of tissues, whereas in Cockayne syndrome, mutations are described in the majority in Cockayne syndrome group B protein (CSB) involved in telomere length maintenance with TRF1 and TRF2 [21]. Prematurely shortened telomeres can also lead to loss of telomere capping function, which in turn can cause telomeres to behave as DNA breaks and undergo recombination to promote genomic instability leading to cell death or tumorigenesis [22]. Telomerase defects have also been associated with

S. cerevisiae	*S. pombe*	*H. sapiens*	Regulates telomerase	Major functions at telomeres	Refs
Rap1	Rap1	RAP1	–	Inhibits NHEJ, plays a role in telomere heterochromatin. ***Only in Sc***: binds telomeric dsDNA	[60, 61, 74–76]
Rif1	Rif1	RIF1	***Sc*** & ***Sp***: –	Only in ***Sc***: checkpoint inhibition at telomeres (with ***Rif2***). ***Hs***: not associated with wild-type telomeres (only at dysfunctional ends to promote NHEJ)	[60, 61, 77, 78]
-np-	Taz1	TRF1 TRF2	–	***Sp*** & ***Hs***: binds telomeric dsDNA, inhibits NHEJ, role in telomere heterochromatin	[58, 61, 79, 80]
YKu70/Yku80	Pku70/Pku80	KU70/KU86	Only in ***Sc***: +	Protects ends from being degraded. Affects subtelomeric heterochromatin	[81–84]
-np-	Poz1	TIN2	–	***Sp*** & ***Hs***: required for the "bridged organization"	[70, 71]
-np-	Tpz1	TPP1	+	***Hs***: telomerase recruitment & processivity. ***Sp*** & ***Hs***: telomerase activation	[68, 69, 85]
-np-	Ccq1	*-np-*	+	***Sp***: telomerase recruitment	[64–67]
-np-	Pot1	POT1	No role	Binds telomeric ssDNA, end protection	[62, 63]
Cdc13 (Est4)	*-np-*	*-np-*	+	Binds telomeric ssDNA, end protection and telomerase recruitment	[38, 86–88]
Stn1/Ten1	Stn1/Ten1	STN1/TEN1	–	End protection. Terminator of telomerase activity/ lagging-strand replication of telomere DNA. Form CST complex with Cdc13 in ***Sc*** and ***CTC1*** in ***Hs***	[69, 89, 90]

dsDNA: double-stranded DNA. -np-: not present. ssDNA: single-stranded DNA.
"+" means positively regulates telomerase. "–" means negatively regulates telomerase.

Table 1.
Major telomere proteins and functions of ***S. cerevisiae, S. pombe,*** *and* ***H. sapiens—Sc: S. cerevisiae, Sp: S. pombe, Hs: H. sapiens,*** *NHEJ: non-homologous-end-joining.*

cancer, and telomerase promoter mutations are prevalent in multiple cancer types [23], where the presence of telomerase activity allows extended growth of cancer cells [24].

S. cerevisiae	*S. pombe*	*H. sapiens*	
Est1	Est1	EST1	Regulatory subunit of the telomerase [91–93]
Est2	Trt1	hTERT	Catalytic subunit of the telomerase [93]
Est3	*-np-*	*-np-*	Regulatory subunit **only in *Sc*** [94]
Tlc1	TER1	hTR (hTERC)	Telomerase RNA [7, 95–97]

Table 2.
Major telomerase components of S. cerevisiae, S. pombe, and H. sapiens.

Telomeres are also able to convert adjacent chromatin into heterochromatin, a compact and transcriptionally silent form of chromatin with important roles in chromosome biology, for example, in development and mammalian female X-chromosome inactivation and development of cancers [25–27]. This function, the inhibition of the DDR, and how the cells distinguish a DSB from a telomere or heterochromatin regulation at the chromosome ends are still not fully understood. Elucidation of these mechanisms will provide an understanding of some of the molecular mechanisms associated with aging and the initiation of tumorigenesis.

1.2 Telomere complexes from yeasts to humans

The yeast model systems have the advantage of a small eukaryotic genome that can be easily altered due to high levels of homologous recombination [28–31]. The terminal telomere repeat tracts of *Saccharomyces cerevisiae* and *Schizosaccharomyces pombe* follow the scale of these smaller genomes with lengths of ~0.3 kb [12, 32]. In contrast, telomeres size in humans is between 10 and 30 kb, but with measurement limitations due to a highly variable length either at the time of birth [33, 34], in different tissues [35] or among others between women and men [36]. Therefore, yeasts are significant model organisms for dissecting the molecular genetics of basic telomere biology (reviewed in [37, 38]).

Telomeres have also been studied in several budding yeasts somewhat related to *S. cerevisiae* [39, 40]. A striking difference between *S. cerevisiae* and these organisms is that while *S. cerevisiae* has a short heterogeneous telomere repeat sequence (abbreviated as TG_{1-3}), several other species of *Saccharomyces*, *Candida*, and *Kluyveromyces lactis* have longer, homogeneous repeats [39, 40]. These longer repeats allowed the cloning of the gene for telomerase RNA in *K. lactis* and *Saccharomyces castellii* (now referred to as *Naumovozyma castellii*) and the study of highly related telomere binding proteins that co-evolved with the different telomere sequences [39, 41–45]. Work in *N. castellii* showed that the Rap1 protein of this yeast can bind to both the double-stranded telomere repeats and the junction between the double and single strands DNA, suggesting a level of coordination between Rap1 and the single-strand telomere repeat binding protein Cdc13 [46, 47]. *N. castellii* may also provide insight into the regulation of telomerase synthesis of the 3′ overhang as this yeast synthesizes a specific 70 nucleotides 3′ overhang in late S-phase [48].

Mutation of the telomerase RNA in *K. lactis* allowed the synthesis of mutant telomere repeats which revealed surprising aspects of telomerase-dependent and telomerase-independent telomere elongation. Some mutant telomere repeats gave rise to hyper-extended telomeres in a phenomenon called "runaway telomere elongation," which was thought to be related to the loss of telomere binding proteins that negatively regulate telomerase [42, 49]. Similar hyper-elongation phenotypes are known in the more "popular" yeasts, *S. cerevisiae* and *S. pombe*, caused by loss of

Rif1 and Rif2 from budding yeast or Taz1 or Rap1 from *S. pombe* (**Table 1**). *K. lactis* mutant telomere repeats have also been used to follow a telomerase-independent mode of recombinational telomere elongation (RTE). In this model, shortened telomeres become highly recombinogenic, and the formation of double-stranded circular DNAs made of telomere repeats (t-circles) allows a "rolling circle" replication mechanism to produce telomere repeats that can be incorporated into short chromosomal telomeres [45, 50] This model is supported by further work indicating that a single t-circle may be a sufficient source for the elongation of all chromosomal telomeres [51, 52]. Results from these studies indicate that *K. lactis* telomeres may become very short prior to RTE and provide a model for human cancers that replicate telomeres by a recombinational mechanism termed ALT for alternative lengthening of telomeres [53]. Therefore, these budding yeasts have provided insights into the mechanisms of telomerase-dependent and -independent telomere elongation. However, both of them lack a rapidly inducible telomere formation system, which has the potential to follow some of the unique processes observed in these yeasts in real time.

The fission yeast *S. pombe* is significantly different from *S. cerevisiae*, *K. lactis*, and *N. castellii*, having diverged from the budding yeasts early in fungal evolution [54]. A consequence of this divergence is that *S. pombe* has conserved many features of telomere protein and chromatin structure that were lost from the budding yeasts. *S. pombe* has the same genome size as *S. cerevisiae* but only three chromosomes compared to 16 for the same size genome, and early studies confirmed that *S. pombe* was amenable to the same molecular genetic manipulations as *S. cerevisiae* [28] including telomere formation [32, 55], but had differences in telomere repeat sequence and the structure of the repeated subtelomeric elements adjacent to chromosome ends [32]. The human telomere complex, also called "shelterin," is more similar to the telomere complex of *S. pombe* than to *S. cerevisiae* (**Tables 1** and **2**; **Figure 1**) (reviewed in [56, 57]). Rap1 does not bind directly to the double-strand telomeric DNA in humans or *S. pombe,* but is recruited to the telomeres by its interaction with double-stranded DNA binding proteins, that is, Taz1 in *S. pombe* and TRF1 and TRF2 in humans. In *S. pombe*, Taz1 and Rap1 inhibit NHEJ, playing a role in the silencing of subtelomere regions and negatively regulating telomerase [58–61]. At the single strand, Pot1 protects the 3′ overhang, as in humans, and is bound by Tpz1 [62, 63]. Tpz1, the ortholog of human TPP1, plays central roles in telomerase recruitment (via its interaction with Ccq1 [64–67]), telomerase activation (via its interaction with the catalytic subunit of the telomerase [68]) and the termination of telomere elongation (via its SUMOylation [69]). Poz1, the ortholog of human TIN2, links the double-strand DNA binding complex of Taz1-Rap1 and Rif1 to the single-strand DNA binding complex of Pot1-Tpz1-Ccq1 (**Figure 1**). The entire complex negatively regulates telomerase as loss of Poz1 or Rap1 results in elongation of the terminal telomere repeat sequence tracts [70, 71]. In addition to telomere proteins, *S. pombe* has also conserved heterochromatin-mediated transcriptional silencing via di- and tri-methylation of lysine 9 of histone H3 (H3K9me2 and H3K9me3), which is absent in *S. cerevisiae* [72, 73].

2. Telomere formation systems in yeasts

Inducible systems that form telomeres from short tracts of telomere repeats were first developed in the late 1990s in *S. cerevisiae* [98]. The system modified inducible DSB systems, which consists of the rapid expression of a restriction enzyme or endonuclease allowing a cut at a specific engineered locus (**Figure 2A** and **B**). These systems have variants in many organisms (reviewed in [99]) with relevance for fundamental understanding of DSB repair mechanisms [100] and therapeutic response to DSBs, as in radiotherapies [101]. In yeast, two site-specific endonucleases, HO and I-SceI were used, with a similar kinetics of cleavage [102, 103]. However, I-SceI was preferred in early mammalian work [104].

The telomere formation system integrates telomeric DNA repeats next to an engineered cut site, and the newly exposed telomere tract is converted into a functional, stable telomere without causing growth inhibition (**Figure 2C**) [98, 105–107]. In contrast, an induced DSB in the middle of a chromosome leads to DNA degradation and growth inhibition [108]. A variant of this system involving an inducible recombinase and specific recognition sites was also made to target one specific telomere in budding yeast and artificially shorten it [89, 109–111].

The *S. cerevisiae* telomere formation system proved to be a highly useful tool, but was limited by the biology of this yeast species. By translating this system to *S. pombe,* some of these issues, such as the presence of H3K9me2 and 3 modifications and a high level of conservation of the telomeric complexes with humans were addressed (**Figure 1B** and **C**) [56, 70, 72, 73]. The recent development of the *S. pombe* telomere formation system thus opens new avenues to study telomeric chromatin regulation and telomere formation.

Telomere formation systems allow a real-time study of telomere formation, elongation or heterochromatin spreading from the newly induced end. It also gives us the opportunities to examine the effects of different mutations in telomere proteins or to test the protein requirements for how telomeres are distinguished from DSBs. This non-exhaustive list highlights the significant advantages of these systems compared to steady-state experiments, as studies introducing mutations into cells with existing telomeres can only monitor telomeres after they have reached their equilibrium state. Formation systems can also study the initiation of mechanisms associated with telomeres, such as heterochromatinization of the nearby sequences or DNA damage checkpoint inhibition. Technically, these systems allow multiple experiments such as ChIP for protein recruitment at the break, Southern blotting to follow the *de novo* telomere elongation or DNA degradation, cell morphology observation (as large budded cells in *S. cerevisiae* or elongated cells in *S. pombe*—characteristic of a G2/M arrest), or western blotting for Chk1 phosphorylation (checkpoint activation).

2.1 Inducible telomere systems in *S. cerevisiae*

The first inducible telomere formation system in *S. cerevisiae* used the HO endonuclease placed under the control of a galactose-inducible promoter (in a *MATa-inc* allele strain which contains a point mutation to avoid an unwanted HO cut at the *MAT* locus [112]) and to the HO recognition sequence placed next to an 80 bp tract of telomere repeats (**Figure 3A**) [98].

The telomere cassette containing the *ADE2* marker, the HO recognition site, and the *LYS2* marker, was inserted at the *ADH4* locus at the left arm of the chromosome VII (VII-L) at 20 kb from the telomere end (this 20 kb contains no essential genes) (**Figure 3A**). The *LYS2* gene here serves as a marker for loss of the non-essential distal DNA and indirectly measure the HO cut efficiency by comparing Lys+ and Lys− cells.

The Diede and Gottsheing studies [98, 113] highlighted the efficacy of the system: a rapid and efficient cutting exposed a new stable telomere end that was elongated by telomerase. They also described a cell cycle-regulated elongation and the requirement of the DNA primase and DNA polymerases α and δ for the *de novo* telomere elongation [98]. Additionally, they also revealed the involvement of the MRX (Mre11-Rad50-Xrs2) complex and its exonuclease activity at the new telomere end for its elongation and Cdc13 loading [113]. It is worth noting that short formed

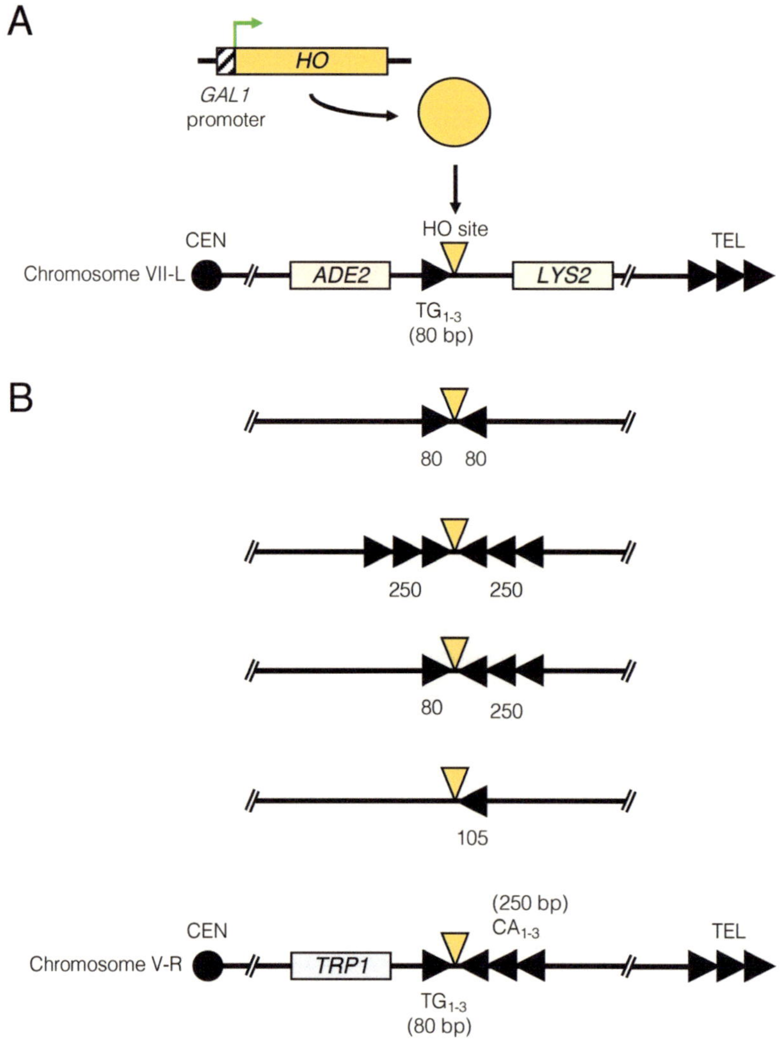

Figure 3.
Inducible telomere formation system in S. cerevisiae (A) schematic representation of the induction of the HO endonuclease (under the control of GAL1 promoter) in presence of galactose and modified chromosome VII-L containing the telomere cassette. The telomere cassette containing the ADE2 marker, the HO recognition site, and the LYS2 marker was inserted at the ADH4 locus at 20 kb from the telomere end (this 20 kb contains no essential genes) [98]. This insertion placed the HO recognition site at 13 kb from the distal telomere end. CEN: centromere. TEL: telomere (terminal black triangles). TG_{1-3}: telomere repeats sequence. (B) Variants of telomeres cassettes. The telomere cassette at the chromosome V-R containing the TRP1 marker was inserted at the YER188W locus (there are no essential genes from this locus to the chromosome end) [77]. Numbers represent the telomere tract length in base pairs (bp). CA_{1-3}: telomere repeats on the other side of the break in the opposite orientation represented in dark red. Modified and adapted from Ribeyre and Shore [77].

telomeres in different mutant strains that have a functional chromosome ends do not require MRX for elongation [114], highlighting how a broken end forming a telomere may differ from an existing end that requires elongation. Finally, they showed that the interaction of Ku (YKu70/YKu80) and telomerase RNA (*TLC1*) promotes telomere addition at the newly formed chromosome end [115].

Michelson et al. showed that even if the telomere formation system has the characteristics of a DSB, the cells respond differently when the DSB is next to a telomeric tract [116]. The *de novo* telomere end and degrading DNA fragment (**Figure 1C**) does not induce a normal checkpoint arrest, giving rise to a "telomere anti-checkpoint" activity.

The Shore lab followed up these results by creating new variants of the systems (**Figure 3B**) with some variations of the telomere tract length, or the addition of inverted telomere tract directly after the HO cut site (distal DNA fragment) [77]. Similarly, a telomere cassette was added at a different locus in the right arm of the chromosome V (V-R) with *TRP1* as the only proximal marker (**Figure 3B**) [77]. These variants were used to show that the telomere anti-checkpoint activity required the telomere proteins: Rif1 and Rif2. These results imply that a telomeric tract on one side of a break influences the recruitment of DNA damage checkpoint proteins on the other side [77].

An alternative inducible telomere elongation system has been constructed where an existing telomere with a functional chromosome end loses internal telomere repeats, so the resulting telomere tract is much shorter than normal [89, 109–111]. The first one used the site-specific recombinase Flp1 under a galactose-inducible promoter and two FRT (Flp1-Recognition Target) sites inserted at the *ADH4* locus on chromosome VII-L (**Figure 4A**) [109, 110, 117]. The system is based on the Flp1 recombinase excising a tract between two FRT sites and leaving only the distal telomere DNA sequence next to the terminal FRT site. Because the *FLP1* gene is encoded on the *S. cerevisiae* endogenous plasmid called the "2-micron circle," only cir° strains that lack this plasmid can be used with the FRT system. The system was later improved by using the loxP sites instead of FRT, and the Cre-recombinase, allowing its use in any strains, at the chromosome VII-L [111] or chromosome V-R, with an unchanged or short telomere induction (**Figure 4B** and **C**) [89, 118]. The constructs contain 16 inverted Rap1-binding sites, a loxP site, and a terminal telomere tract. Arrays of Rap1 binding sites at this position are considered by the cell to be a part of the terminal telomere tract, so the internal repeat tract distal to the loxP site is shorter than a normal telomere [119–121]. In these systems, after recombination, the remaining terminal tract is about 90 bp or 1/3 of the normal telomere length (**Figure 4B**) [89, 109–111, 117, 118]. These constructs were used to show a cell cycle restriction of telomere elongation in late S phase and a progressive telomerase inhibition upon telomere elongation [109, 110]. Additionally, an early replication of a short telomere was linked to an increase association of telomerase [111, 117] and highlighted two distinct roles of Stn1 in telomere capping and telomerase inhibition [89].

2.2 Inducible telomere systems in *S. pombe*

More recently, the telomere formation system was extended to *S. pombe*. The telomeric complex of the fission yeast has a high level of conservation with humans, both structurally with a "bridged organization" of telomere proteins [70] and in how telomerase is regulated (**Figure 1B** and **C**) (reviewed in [56]). This yeast also exhibits the H3K9me2 and -3 modifications present in multicellular eukaryotes that is absent in *S. cerevisiae* opening new perspectives of translational studies in telomeric chromatin regulations [72, 73].

The building of an efficient system took more time in *S. pombe*, as a rapidly inducible promoter was not available in this system. A system with the HO endonuclease using the *nmt1* promoter had been used but required 24 hours for full induction that would complicate kinetic analysis [122]. Two systems that could rapidly induce a site-specific nuclease that did not cut in the *S. pombe* genome were published in 2011 and 2012, but were inappropriate for a telomere formation system because they either prevented the use of the *ura4⁺* marker, which is important for monitoring telomere-associated heterochromatin [123–125], or special strains [126]. We, therefore, developed a telomere system that was using the anhydrotetracycline (ahTET)-inducible promoter and an I-SceI enzyme with preferred *S. pombe* codons [127] coupled to two NLS (Nuclear Localization Signals) to ensure a rapid

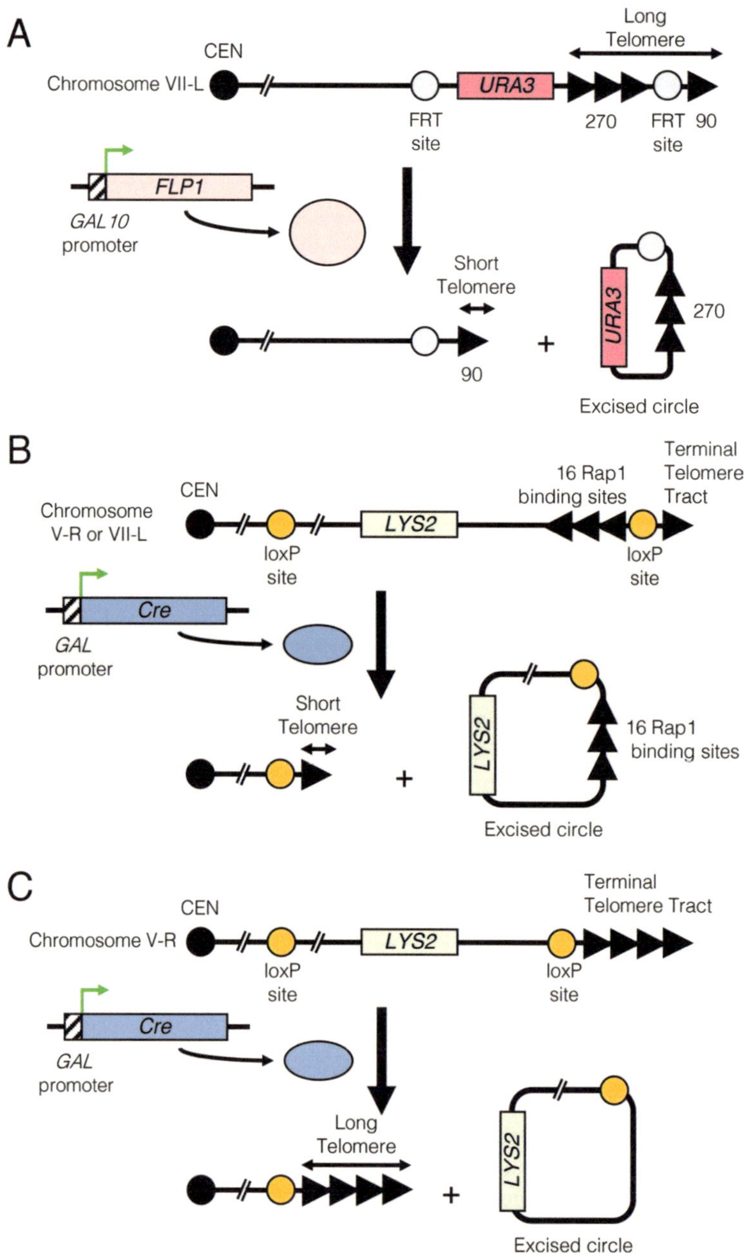

Figure 4.
S. cerevisiae systems to generate a single shortened telomere (A) schematic representation of the inducible recombinase-based system to generate a single shortened telomere at the chromosome VII-L (ADH4 locus). FRT: Flp1-recognition target site. The galactose-mediated induction allows the expression of Flp1 which induced homologous recombination through the FRT sites to generate an excised circle (containing the indicated elements) and a short telomere. Numbers represent the telomere tract length in base pairs (bp). Note that the terminal tract length is variable at individual telomeres due to lengthening and shortening at each cell division, and the numbers presented represent the average of the population [109, 110, 117]. (B) Variant system using loxP site and galactose inducible Cre-recombinase expression at the same locus (ADH4, chromosome VII-L) [111] or at the chromosome V-R end (YER188W locus) [89, 118]. (C) The control chromosome V-R construct that does not excise telomere repeats [89]. Modified and adapted from Marcand et al. and Puglisi et al. [89, 109].

and efficient cut [107] (**Figure 5A**). I-SceI has no endogenous site in the fission yeast genome [128], so the only DSB induced is at the engineered I-SceI site in the telomere formation system (called the proto-telomere cassette in *S. pombe*).

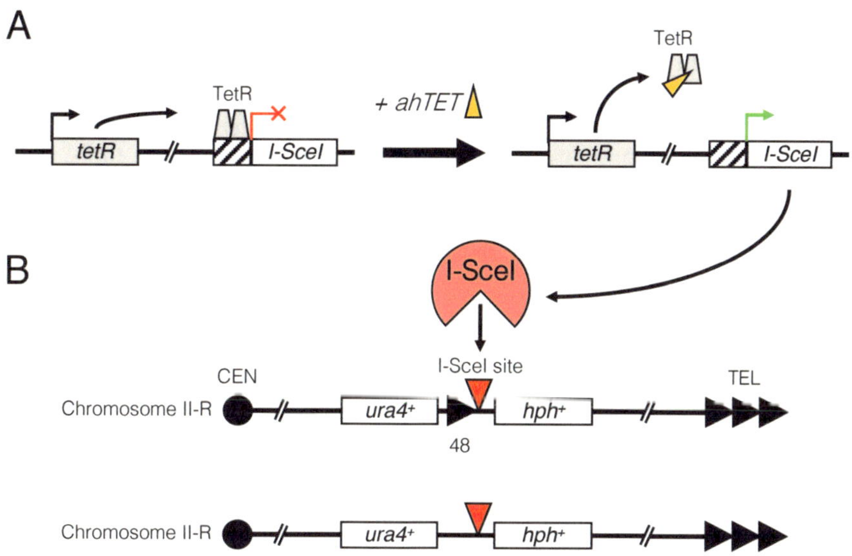

Figure 5.
The S. pombe inducible telomere formation system (A) the I-SceI endonuclease was expressed from a TetR controlled promoter (CaMV35S). In the absence of ahTET (left part), TetR represses the expression of I-SceI, and the addition of ahTET (9uM) into the medium induces expression (right part). (B) The 48 and 0 bp proto-telomeres cassettes are presented here and consist of a ura4⁺ gene followed by either a 48 bp of telomere tract (black triangles) or no telomere sequences for the 0 bp version, the I-SceI cut site (red triangle) and the hygromycin resistance gene (hph⁺) at the distal part of the construct. The cassettes were introduced at the chromosome II-R, 3′ to the gal1⁺ locus. CEN: centromere. TEL: telomere (terminal black triangles). The number represents the telomere tract length in base pairs (bp). Modified and adapted from Wang et al. [107].

Two constructs were then created containing either 48 or 0 bp of telomere repeat sequence, an I-SceI cut site, and two flanking selectable markers: the *ura4⁺* gene and the hygromycin resistance gene (*hph⁺*) (**Figure 5B**) [107]. The cassettes were placed at the right arm of the chromosome II (II-R), 3′ to the *gal1⁺* locus and, as in the *S. cerevisiae* systems, the region between this locus and the chromosome end was dispensable (without essential genes).

The *S. pombe* telomere formation system revealed several novel aspects of telomere function [107]. First, the DSB next to the telomere repeat tract immediately acquires telomere function: the end is immediately stable and not degraded (even in a strain lacking telomerase activity), does not undergo recombination, and the DNA damage response is somehow blocked. Second, the heterochromatin domain associated with telomeres formed in two distinct phases. The first consists of heterochromatin spreading from the telomere toward the interior of the chromosome over ~3 kb in the first cell division (about 5 hours). The newly formed telomere is elongated to wild type lengths during this time by telomerase. Heterochromatin spreading over ~10 kb continues for the next eight generations (population doublings) after the terminal telomere repeat tracts are at their equilibrium lengths. It is worth noting that the *clr4* deletion, which blocks heterochromatin formation, had no effect on telomere elongation [107], indicating that heterochromatin is independent of telomerase-mediated elongation as it is for chromosome end stability as observed in *S. cerevisiae* [129, 130]. In a second phase studied 34–87 population doublings after telomere formation, the established heterochromatin domain was surprisingly dynamic, with significant expansions and contraction of the heterochromatin mark H3K9me2 over the 35 kb domain that was monitored. Thus, different lineages from the same telomere formation event had different heterochromatin domains at different times after formation. These lineages were monitored at the single cell level by integrating the *ade6⁺* colony color marker at different distances from the proto-telomere and plating single cells on inducing medium to follow the spreading of heterochromatin to silence the color marker at different stages of colony growth [107]. A marker close to the newly formed telomere initially showed expression before it was extinguished after several population doublings and remained extinguished as cells continued to grow. In contrast, markers placed further away from the telomere switched between the transcribed and repressed state in subsequent population doublings. These relative kinetics of telomere-associated functions would not have been observable using a telomere formed by cellular transformation, which requires ~30 generations of growth between telomere formation of a single cell to the generation of a sufficient number of cells to analyze the heterochromatin domain.

It is worth noting that this work and many other *S. pombe* heterochromatin studies follow the histone modification H3K9me2, but the H3K9me**3** mark is also associated with heterochromatin. Recent work by the Moazed lab showed that the H3K9me2 mark is associated with very low gene activity, and its conversion to H3K9me3 extinguished detectable transcription [131]. This model can explain the level of H3K9me2 near the newly formed telomere, which peaked about 3–13 kb from the new telomere in different experiments [107]. Presumably, H3K9me3 modification replaced H3K9me2 closer to the new telomere where expression of the *ura4*$^{+}$ and *ade6*$^{+}$ genes was undetectable [107]. Thus, the inducible telomere formation system may also prove useful in studying how the transition between these chromatin marks is regulated.

3. Conclusion

The inducible telomere formation system first developed in *S. cerevisiae* has provided several important results in the understanding of telomere elongation, in the cell cycle regulation of telomerase, in DNA checkpoint inhibition induced by a telomere end, and in the role of specific telomere proteins (**Table 3**) [77, 89, 98, 109, 110, 113, 115–118]. The new *S. pombe* system has allowed one to follow the establishment of telomere-end protection functions and the formation and dynamics of heterochromatin (**Table 3**) [107].

The ability to monitor the relative kinetics of different telomere-associated processes of replication, end protection, and chromatin domain formation has provided insights that could not be obtained using steady-state experiments with established telomeres. As telomere dysfunctions contribute to genomic instability or chromosome aberrations in human cancers, aging disorders, or specific pathologies such as dyskeratosis congenita [10, 19–21, 23, 24], telomere formation studies have the potential to identify defects related to these diseases. Heterochromatin also plays an important role in chromosome biology [25–27], and inducible telomeres and heterochromatin domains may shed light on these metazoan processes.

A key component of these telomere formation systems is the rapidly inducible DSB to expose the new physical chromosome end with adjacent telomere repeats. Inducible DSB systems using I-SceI or CRISPR/Cas9 have been used in mammalian cells to reveal differences in DNA repair at internal sites compared to subtelomeric ones [132–137]. An issue with the CRISPR/Cas9 approach is that it is a two-component system consisting of a guide RNA to target the endonuclease to a precise site, so how rapidly inducible this system is compared to the single component I-SceI nuclease needs to be carefully tested. Using I-SceI, an inducible telomere system in mouse cells to measure telomere repeat addition over 48 hours was developed [105]. While not used kinetically, the system allowed the assay of the effect of the DNA damage kinase ATM on the addition of new telomere repeats by telomerase.

Organism	System description	Used to show
S. cerevisiae	• HO endonuclease • Galactose inducible promoter • Specific cut site at chromosome VII-L or V-R to create a new telomere end	• New telomere stable and elongation by telomerase [98]. • Telomere elongation is cell cycle regulated [98]. • DNA primase and DNA polymerases α and δ are required of the *de novo* telomere elongation [98]. • Involvement of the MRX complex and its exonuclease activity at the new telomere end for its elongation and the Cdc13 loading [113]. • Ku complex interacts with *TLC1* (telomerase RNA) to promote telomere addition at the newly formed chromosome end [115]. • The *de novo* telomere end and degrading DNA fragment does not induce a checkpoint arrest [116].
S. cerevisiae	• FLP1 recombinase • Galactose inducible promoter • FRT sites inserted at chromosome VII-L to artificially shorten a telomere end	• Progressive telomerase inhibition upon telomere elongation [109]. • Cell cycle restriction of telomere elongation in late S phase [110]. • Telomerase is preferentially associated with short telomeres [117].
S. cerevisiae	• Cre recombinase • Galactose inducible promoter • LoxP sites inserted at chromosome VII-L or V-R to artificially shorten a telomere end	• Early replication of a short telomere linked to an increased association of telomerase [111]. • Two distinct roles of Stn1 in telomere capping and telomerase inhibition [89].
S. pombe	• I-SceI endonuclease • Tetracycline (TetR) controlled promoter (CaMV35S) • Specific cut site at chromosome II-R to create a new telomere end	• New telomere-end stable and elongated by telomerase [107]. • Gradual heterochromatin formation which remains dynamic after the new end reaches its equilibrium length [107].

Table 3.
Summary of telomere formation systems in different yeast species.

The events discovered in the *S. cerevisiae* and *S. pombe* telomere formation systems will provide important models for testing in this and other mammalian telomere formation systems.

Acknowledgements

This work was supported by grants from the National Science Foundation and the National Institutes of Health (both USA) to KWR. The funding agencies had no influence on the content of this chapter.

Conflict of interest

The authors declare no conflicts of interest.

Author details

Julien Audry[1*] and Kurt W. Runge[1,2]

1 Department of Inflammation and Immunity, Lerner Research Institute, Cleveland Clinic Foundation, Cleveland, Ohio, USA

2 Department of Genetics and Genome Sciences, Case Western Reserve University, Cleveland, Ohio, USA

*Address all correspondence to: audryj@ccf.org

References

[1] Muller HJ. The remaking of chromosomes. The Collect Net. 1938;**3**:181-198

[2] McClintock B. The stability of broken ends of chromosomes in Zea Mays. Genetics. 1941;**26**(2):234-282

[3] Watson JD. Origin of concatemeric T7 DNA. Nature New Biology. 1972;**239**(94):197-201

[4] Olovnikov AM. A theory of marginotomy. The incomplete copying of template margin in enzymic synthesis of polynucleotides and biological significance of the phenomenon. Journal of Theoretical Biology. 1973;**41**(1):181-190

[5] Greider CW, Blackburn EH. I dentification of a specific telomere terminal transferase activity in tetrahymena extracts. Cell. 1985;**43**(2):405-413

[6] Hemann MT, Strong MA, Hao LY, Greider CW. The shortest telomere, not average telomere length, is critical for cell viability and chromosome stability. Cell. 2001;**107**(1):67-77

[7] Collins K, Mitchell JR. Telomerase in the human organism. Oncogene. 2002;**21**(4):564-579

[8] Cosme-Blanco W, Shen M-F, Lazar AJF, Pathak S, Lozano G, Multani AS, et al. Telomere dysfunction suppresses spontaneous tumorigenesis *in vivo* by initiating p53-dependent cellular senescence. EMBO Reports. 2007;**8**(5):497-503

[9] Feldser DM, Greider CW. Short telomeres limit tumor progression *in vivo* by inducing senescence. Cancer Cell. 2007;**11**(5):461-469

[10] Cesare AJ, Karlseder J. A three-state model of telomere control over human proliferative boundaries. Current Opinion in Cell Biology. 2012;**24**(6):731-738

[11] Murray AW, Schultes NP, Szostak JW. Chromosome length controls mitotic chromosome segregation in yeast. Cell. 1986;**45**(4):529-536

[12] Wang SS, Zakian VA. Sequencing of *Saccharomyces* telomeres cloned using T4 DNA polymerase reveals two domains. Molecular and Cellular Biology. 1990;**10**(8):4415-4419

[13] de Lange T, Shiue L, Myers RM, Cox DR, Naylor SL, Killery AM, et al. Structure and variability of human chromosome ends. Molecular and Cellular Biology. 1990;**10**(2):518-527

[14] Sfeir A, de Lange T. Removal of shelterin reveals the telomere end-protection problem. Science. 2012;**336**(6081):593-597

[15] de Lange T. Protection of mammalian telomeres. Oncogene. 2002;**21**(4):532-540

[16] Hao L-Y, Armanios M, Strong MA, Karim B, Feldser DM, Huso D, et al. Short telomeres, even in the presence of telomerase, limit tissue renewal capacity. Cell. 2005;**123**(6):1121-1131

[17] Tomás-Loba A, Flores I, Fernández-Marcos PJ, Cayuela ML, Maraver A, Tejera A, et al. Telomerase reverse transcriptase delays aging in cancer-resistant mice. Cell. 2008;**135**(4):609-622

[18] Fasching CL. Telomere length measurement as a clinical biomarker of aging and disease. Critical Reviews in Clinical Laboratory Sciences. 2018;**55**(7):443-465

[19] Opresko PL, Shay JW. Telomere-associated aging disorders. Ageing Research Reviews. 2017;**33**:52-66

[20] Kirwan M, Dokal I. Dyskeratosis congenita, stem cells and telomeres. Biochimica et Biophysica Acta. 2009;**1792**(4):371-379

[21] Turner KJ, Vasu V, Griffin DK. Telomere biology and human phenotype. Cell. 2019;**8**(1):pii:E73

[22] Blasco MA, Lee HW, Hande MP, Samper E, Lansdorp PM, DePinho RA, et al. Telomere shortening and tumor formation by mouse cells lacking telomerase RNA. Cell. 1997;**91**(1):25-34

[23] Heidenreich B, Rachakonda PS, Hemminki K, Kumar R. TERT promoter mutations in cancer development. Current Opinion in Genetics & Development. 2014;**24**:30-37

[24] Shay JW, Wright WE. Role of telomeres and telomerase in cancer. Seminars in Cancer Biology. 2011;**21**(6):349-353

[25] Chaligné R, Heard E. X-chromosome inactivation in development and cancer. FEBS Letters. 2014;**588**(15):2514-2522

[26] Lunyak VV, Rosenfeld MG. Epigenetic regulation of stem cell fate. Human Molecular Genetics. 2008;**17**(R1):R28-R36

[27] Sharma S, Kelly TK, Jones PA. Epigenetics in cancer. Carcinogenesis. 2010;**31**(1):27-36

[28] Simanis V. The molecular biology of *Schizosaccharomyces pombe*: Genetics, genomics and beyond. Journal of Cell Science. 2004;**117**(17):3712-3712

[29] Hoekstra MF. In: Jones EW, Pringle JR, Broach JR, editors. The Molecular and Cellular Biology of the Yeast Saccharomyces. Vol. 1: Genome Dynamics, Protein Synthesis, and Energetics, Trends in Genetics. Cold Spring Harbor USA, New York: Laboratory Press (Monograph 21A); 1992

[30] Newman A. In: Jones EW, Pringle JR, Broach JR, editors. The Molecular and Cellular Biology of the Yeast Saccharomyces. Vol. 2: Gene Expression, Trends in Cell Biology. Cold Spring Harbor USA, New York: Laboratory Press (Monograph 21B); 1994

[31] Fantes PA. In: Jones EW, Pringle JR, Broach JR, editors. The Molecular and Cellular Biology of the Yeast Saccharomyces. Vol. 3: Cell Cycle and Cell Biology, Trends in Genetics. Cold Spring Harbor USA, New York: Laboratory Press (Monograph 21C); 1998

[32] Sugawara NF, thesis PD. DNA Sequences at the Telomeres of the Fission Yeast *S. pombe*. Cambridge, Massachusetts, USA: Harvard University; 1988

[33] Okuda K, Bardeguez A, Gardner JP, Rodriguez P, Ganesh V, Kimura M, et al. Telomere length in the newborn. Pediatric Research. 2002;**52**(3):377-381

[34] Vasu V, Turner KJ, George S, Greenall J, Slijepcevic P, Griffin DK. Preterm infants have significantly longer telomeres than their term born counterparts. PLoS One. 2017;**12**(6):e0180082

[35] Friedrich U, Griese E, Schwab M, Fritz P, Thon K, Klotz U. Telomere length in different tissues of elderly patients. Mechanisms of Ageing and Development. 2000;**119**(3):89-99

[36] Nordfjäll K, Eliasson M, Stegmayr B, Melander O, Nilsson P, Roos G. Telomere length is associated with obesity parameters but with a gender difference. Obesity. 2008;**16**(12):2682-2689

[37] Bianchi A, Shore D. How telomerase reaches its end: Mechanism of telomerase regulation by the telomeric complex. Molecular Cell. 2008;**31**(2):153-165

[38] Churikov D, Corda Y, Luciano P, Géli V. Cdc13 at a crossroads of telomerase action. Frontiers in Oncology. 2013;**3**:39

[39] McEachern MJ, Blackburn EH. A conserved sequence motif within the exceptionally diverse telomeric sequences of budding yeasts. Proceedings of the National Academy of Sciences of the United States of America. 1994;**91**(8):3453-3457

[40] Cohn M, McEachern MJ, Blackburn EH. Telomeric sequence diversity within the genus *Saccharomyces*. Current Genetics. 1998;**33**(2):83-91

[41] Cohn M, Blackburn EH. Telomerase in yeast. Science. 1995;**269**(5222):396-400

[42] Krauskopf A, Blackburn EH. Control of telomere growth by interactions of RAP1 with the most distal telomeric repeats. Nature. 1996;**383**(6598):354-357

[43] Gustafsson C, Rhodin Edsö J, Cohn M. Rap1 binds single-stranded DNA at telomeric double- and single-stranded junctions and competes with Cdc13 protein. The Journal of Biological Chemistry. 2011;**286**(52):45174-45185

[44] Rhodin Edsö J, Tati R, Cohn M. Highly sequence-specific binding is retained within the DNA-binding domain of the *Saccharomyces castellii* Cdc13 telomere-binding protein. FEMS Yeast Research. 2008;**8**(8):1289-1302

[45] Basenko EY, Cesare AJ, Iyer S, Griffith JD, McEachern MJ. Telomeric circles are abundant in the stn1-M1 mutant that maintains its telomeres through recombination. Nucleic Acids Research. 2010;**38**(1):182-189

[46] Runnberg R, Narayanan S, Cohn M. Rap1 and Cdc13 have complementary roles in preventing exonucleolytic degradation of telomere 5′ ends. Scientific Reports. 2017;**7**(1):8729

[47] Rhodin Edsö J, Gustafsson C, Cohn M. Single- and double-stranded DNA binding proteins act in concert to conserve a telomeric DNA core sequence. Genome Integrity. 2011;**2**(1):2

[48] Fridholm H, Astromskas E, Cohn M. Telomerase-dependent generation of 70-nt-long telomeric single-stranded 3′ overhangs in yeast. Nucleic Acids Research. 2013;**41**(1):242-252

[49] McEachern MJ, Blackburn EH. Runaway telomere elongation caused by telomerase RNA gene mutations. Nature. 1995;**376**(6539):403-409

[50] McEachern MJ, Iyer S. Short telomeres in yeast are highly recombinogenic. Molecular Cell. 2001;**7**(4):695-704

[51] Natarajan S, Groff-Vindman C, McEachern MJ. Factors influencing the recombinational expansion and spread of telomeric tandem arrays in *Kluyveromyces lactis*. Eukaryotic Cell. 2003;**2**(5):1115-1127

[52] Natarajan S, McEachern MJ. Recombinational telomere elongation promoted by DNA circles. Molecular and Cellular Biology. 2002;**22**(13):4512-4521

[53] Xu J, McEachern MJ. Long telomeres produced by telomerase-resistant recombination are established from a single source and are subject to extreme sequence scrambling. PLoS Genetics. 2012;**8**(11):e1003017

[54] Hayles J, Nurse P. Introduction to fission yeast as a model system. Cold Spring Harbor Protocols. 2018;**2018**(5):pdb.top079749

[55] Guerrini AM, Ascenzioni F, Tribioli C, Donini P. Transformation

of *Saccharomyces cerevisiae* and *Schizosaccharomyces pombe* with linear plasmids containing 2 micron sequences. The EMBO Journal. 1985;**4**(6):1569-1573

[56] Nandakumar J, Cech TR. Finding the end: Recruitment of telomerase to telomeres. Nature Reviews Molecular Cell Biology. 2013;**14**(2):69-82

[57] de Lange T. Shelterin-mediated telomere protection. Annual Review of Genetics. 2018;**52**:223-247

[58] Ferreira MG, Cooper JP. The fission yeast Taz1 protein protects chromosomes from Ku-dependent end-to-end fusions. Molecular Cell. 2001;**7**(1):55-63

[59] Miller KM, Rog O, Cooper JP. Semi-conservative DNA replication through telomeres requires Taz1. Nature. 2006;**440**(7085):824-828

[60] Kanoh J, Ishikawa F. spRap1 and spRif1, recruited to telomeres by Taz1, are essential for telomere function in fission yeast. Current Biology. 2001;**11**(20):1624-1630

[61] Miller KM, Ferreira MG, Cooper JP. Taz1, Rap1 and Rif1 act both interdependently and independently to maintain telomeres. The EMBO Journal. 2005;**24**(17):3128-3135

[62] Baumann P. Pot1, the putative telomere end-binding protein in fission yeast and humans. Science. 2001;**292**(5519):1171-1175

[63] Denchi EL, de Lange T. Protection of telomeres through independent control of ATM and ATR by TRF2 and POT1. Nature. 2007;**448**(7157):1068-1071

[64] Moser BA, Chang Y-T, Kosti J, Nakamura TM. Tel1ATM and Rad3ATR kinases promote Ccq1-Est1 interaction to maintain telomeres in fission yeast. Nature Structural & Molecular Biology. 2011;**18**(12):1408-1413

[65] Chang Y-T, Moser BA, Nakamura TM. Fission yeast shelterin regulates DNA polymerases and Rad3ATR kinase to limit telomere extension. PLoS Genetics. 2013;**9**(11):e1003936

[66] Moser BA, Subramanian L, Khair L, Chang Y-T, Nakamura TM. Fission yeast Tel1(ATM) and Rad3(ATR) promote telomere protection and telomerase recruitment. PLoS Genetics. 2009;**5**(8):e1000622

[67] Webb CJ, Zakian VA. *Schizosaccharomyces pombe* Ccq1 and TER1 bind the 14-3-3-like domain of Est1, which promotes and stabilizes telomerase-telomere association. Genes & Development. 2012;**26**(1):82-91

[68] Armstrong CA, Pearson SR, Amelina H, Moiseeva V, Tomita K. Telomerase activation after recruitment in fission yeast. Current Biology. 2014;**24**(17):2006-2011

[69] Garg M, Gurung RL, Mansoubi S, Ahmed JO, Dave A, Watts FZ, et al. Tpz1TPP1 SUMOylation reveals evolutionary conservation of SUMO-dependent Stn1 telomere association. EMBO Reports. 2014;**15**(8):871-877

[70] Kim J-K, Liu J, Hu X, Yu C, Roskamp K, Sankaran B, et al. Structural basis for Shelterin bridge assembly. Molecular Cell. 2017;**68**(4):698-714.e5

[71] Harland JL, Chang Y-T, Moser BA, Nakamura TM. Tpz1-Ccq1 and Tpz1-Poz1 interactions within fission yeast shelterin modulate Ccq1 Thr93 phosphorylation and telomerase recruitment. PLoS Genetics. 2014;**10**(10):e1004708

[72] Allshire RC, Ekwall K. Epigenetic regulation of chromatin states in *Schizosaccharomyces pombe*. Cold

Spring Harbor Perspectives in Biology. 2015;**7**(7):a018770

[73] Martienssen R, Moazed D. RNAi and heterochromatin assembly. Cold Spring Harbor Perspectives in Biology. 2015;**7**(8):a019323

[74] Li B, de Lange T. Rap1 affects the length and heterogeneity of human telomeres. Molecular Biology of the Cell. 2003;**14**(12):5060-5068

[75] Pardo B, Marcand S. Rap1 prevents telomere fusions by nonhomologous end joining. The EMBO Journal. 2005;**24**(17):3117-3127

[76] Wahlin J. *Saccharomyces cerevisiae* RAP1 binds to telomeric sequences with spatial flexibility. Nucleic Acids Research. 2000;**28**(12):2292-2301

[77] Ribeyre C, Shore D. Anticheckpoint pathways at telomeres in yeast. Nature Structural & Molecular Biology. 2012;**19**(3):307-313

[78] Silverman J. Human Rif1, ortholog of a yeast telomeric protein, is regulated by ATM and 53BP1 and functions in the S-phase checkpoint. Genes & Development. 2004;**18**(17):2108-2119

[79] Dehé P-M, Rog O, Ferreira MG, Greenwood J, Cooper JP. Taz1 enforces cell-cycle regulation of telomere synthesis. Molecular Cell. 2012;**46**(6):797-808

[80] Deng Z, Norseen J, Wiedmer A, Riethman H, Lieberman PM. TERRA RNA binding to TRF2 facilitates heterochromatin formation and ORC recruitment at telomeres. Molecular Cell. 2009;**35**(4):403-413

[81] Roy R, Meier B, McAinsh AD, Feldmann HM, Jackson SP. Separation-of-function mutants of yeast Ku80 reveal a Yku80p-Sir4p interaction involved in Telomeric silencing. The Journal of Biological Chemistry. 2004;**279**(1):86-94

[82] Zappulla DC, Goodrich KJ, Arthur JR, Gurski LA, Denham EM, Stellwagen AE, et al. Ku can contribute to telomere lengthening in yeast at multiple positions in the telomerase RNP. RNA. 2011;**17**(2):298-311

[83] Celli GB, Denchi EL, de Lange T. Ku70 stimulates fusion of dysfunctional telomeres yet protects chromosome ends from homologous recombination. Nature Cell Biology. 2006;**8**(8):885-890

[84] Rai R, Zheng H, He H, Luo Y, Multani A, Carpenter PB, et al. The function of classical and alternative non-homologous end-joining pathways in the fusion of dysfunctional telomeres. The EMBO Journal. 2010;**29**(15):2598-2610

[85] Nandakumar J, Bell CF, Weidenfeld I, Zaug AJ, Leinwand LA, Cech TR. The TEL patch of telomere protein TPP1 mediates telomerase recruitment and processivity. Nature. 2012;**492**(7428):285-289

[86] Pennock E, Buckley K, Lundblad V. Cdc13 delivers separate complexes to the telomere for end protection and replication. Cell. 2001;**104**(3):387-396

[87] Li S, Makovets S, Matsuguchi T, Blethrow JD, Shokat KM, Blackburn EH. Cdk1-dependent phosphorylation of Cdc13 coordinates telomere elongation during cell-cycle progression. Cell. 2009;**136**(1):50-61

[88] Hang LE, Liu X, Cheung I, Yang Y, Zhao X. SUMOylation regulates telomere length homeostasis by targeting Cdc13. Nature Structural & Molecular Biology. 2011;**18**(8):920-926

[89] Puglisi A, Bianchi A, Lemmens L, Damay P, Shore D. Distinct roles for yeast Stn1 in telomere capping and telomerase inhibition. The EMBO Journal. 2008;**27**(17):2328-2339

[90] Chen L-Y, Redon S, Lingner J. The human CST complex is a terminator

of telomerase activity. Nature. 2012;**488**(7412):540-544

[91] Reichenbach P, Höss M, Azzalin CM, Nabholz M, Bucher P, Lingner J. A human homolog of yeast Est1 associates with telomerase and uncaps chromosome ends when overexpressed. Current Biology. 2003;**13**(7):568-574

[92] Qi H, Zakian VA. The *Saccharomyces* telomere-binding protein Cdc13p interacts with both the catalytic subunit of DNA polymerase alpha and the telomerase-associated est1 protein. Genes & Development. 2000;**14**(14):1777-1788

[93] Nakamura TM. Telomerase catalytic subunit homologs from fission yeast and human. Science. 1997;**277**(5328):955-959

[94] Hughes TR, Evans SK, Weilbaecher RG, Lundblad V. The Est3 protein is a subunit of yeast telomerase. Current Biology. 2000;**10**(13):809-812

[95] Singer M, Gottschling D. TLC1: Template RNA component of *Saccharomyces cerevisiae* telomerase. Science. 1994;**266**(5184):404-409

[96] Webb CJ, Zakian VA. Identification and characterization of the *Schizosaccharomyces pombe* TER1 telomerase RNA. Nature Structural & Molecular Biology. 2008;**15**(1):34-42

[97] Leonardi J, Box JA, Bunch JT, Baumann P. TER1, the RNA subunit of fission yeast telomerase. Nature Structural & Molecular Biology. 2008;**15**(1):26-33

[98] Diede SJ, Gottschling DE. Telomerase-mediated telomere addition *in vivo* requires DNA primase and DNA polymerases alpha and delta. Cell. 1999;**99**(7):723-733

[99] Jasin M, Haber JE. The democratization of gene editing: Insights from site-specific cleavage and double-strand break repair. DNA Repair. 2016;**44**:6-16

[100] Janssen A, Breuer GA, Brinkman EK, van der Meulen AI, Borden SV, van Steensel B, et al. A single double-strand break system reveals repair dynamics and mechanisms in heterochromatin and euchromatin. Genes & Development. 2016;**30**(14):1645-1657

[101] Mladenova V, Mladenov E, Iliakis G. Novel biological approaches for testing the contributions of single DSBs and DSB clusters to the biological effects of high LET radiation. Frontiers in Oncology. 2016;**6**:163

[102] Plessis A, Perrin A, Haber JE, Dujon B. Site-specific recombination determined by I-SceI, a mitochondrial group I intron-encoded endonuclease expressed in the yeast nucleus. Genetics. 1992;**130**(3):451-460

[103] Haber JE. *In vivo* biochemistry: Physical monitoring of recombination induced by site-specific endonucleases. BioEssays. 1995;**17**(7):609-620

[104] Rouet P, Smih F, Jasin M. Introduction of double-strand breaks into the genome of mouse cells by expression of a rare-cutting endonuclease. Molecular and Cellular Biology. 1994;**14**(12):8096-8106

[105] Lee SS, Bohrson C, Pike AM, Wheelan SJ, Greider CW. ATM kinase is required for telomere elongation in mouse and human cells. Cell Reports. 2015;**13**(8):1623-1632

[106] Ribeyre C, Shore D. Regulation of telomere addition at DNA double-strand breaks. Chromosoma. 2013;**122**(3):159-173

[107] Wang J, Eisenstatt JR, Audry J, Cornelius K, Shaughnessy M, Berkner KL, et al. A heterochromatin domain forms gradually at a new telomere and is

dynamic at stable telomeres. Molecular and Cellular Biology. 2018;**38**(15):pii: e00393-17

[108] Harrison JC, Haber JE. Surviving the breakup: The DNA damage checkpoint. Annual Review of Genetics. 2006;**40**(1):209-235

[109] Marcand S, Brevet V, Gilson E. Progressive cis-inhibition of telomerase upon telomere elongation. The EMBO Journal. 1999;**18**(12):3509-3519

[110] Marcand S, Brevet V, Mann C, Gilson E. Cell cycle restriction of telomere elongation. Current Biology. 2000;**10**(8):487-490

[111] Bianchi A, Shore D. Early replication of short telomeres in budding yeast. Cell. 2007;**128**(6):1051-1062

[112] Sweetser DB, Hough H, Whelden JF, Arbuckle M, Nickoloff JA. Fine-resolution mapping of spontaneous and double-strand break-induced gene conversion tracts in *Saccharomyces cerevisiae* reveals reversible mitotic conversion polarity. Molecular and Cellular Biology. 1994;**14**(6):3863-3875

[113] Diede SJ, Gottschling DE. Exonuclease activity is required for sequence addition and Cdc13p loading at a *de novo* telomere. Current Biology. 2001;**11**(17):1336-1340

[114] Tsukamoto Y, Taggart AKP, Zakian VA. The role of the Mre11-Rad50-Xrs2 complex in telomerase-mediated lengthening of *Saccharomyces cerevisiae* telomeres. Current Biology. 2001;**11**(17):1328-1335

[115] Stellwagen AE, Haimberger ZW, Veatch JR, Gottschling DE. Ku interacts with telomerase RNA to promote telomere addition at native and broken chromosome ends. Genes & Development. 2003;**17**(19):2384-2395

[116] Michelson RJ, Rosenstein S, Weinert T. A telomeric repeat sequence adjacent to a DNA double-stranded break produces an anticheckpoint. Genes & Development. 2005;**19**(21):2546-2559

[117] Sabourin M, Tuzon CT, Zakian VA. Telomerase and Tel1p preferentially associate with short telomeres in *S. cerevisiae*. Molecular Cell. 2007;**27**(4):550-561

[118] Bianchi A, Shore D. Increased association of telomerase with short telomeres in yeast. Genes & Development. 2007;**21**(14):1726-1730

[119] Grossi S, Bianchi A, Damay P, Shore D. Telomere formation by Rap1p binding site arrays reveals end-specific length regulation requirements and active telomeric recombination. Molecular and Cellular Biology. 2001;**21**(23):8117-8128

[120] Marcand S, Gilson E, Shore D. A protein-counting mechanism for telomere length regulation in yeast. Science. 1997;**275**(5302):986-990

[121] Ray A, Runge KW. The yeast telomere length counting machinery is sensitive to sequences at the telomere-nontelomere junction. Molecular and Cellular Biology. 1999;**19**(1):31-45

[122] Du L-L, Nakamura TM, Russell P. Histone modification-dependent and -independent pathways for recruitment of checkpoint protein Crb2 to double-strand breaks. Genes & Development. 2006;**20**(12):1583-1596

[123] Watson AT, Werler P, Carr AM. Regulation of gene expression at the fission yeast *Schizosaccharomyces pombe urg1* locus. Gene. 2011;**484**(1-2):75-85

[124] Nimmo ERR, Cranston G, Allshire RCC. Telomere-associated chromosome breakage in fission yeast results in variegated expression of

adjacent genes. The EMBO Journal. 1994;**13**(16):3801-3811

[125] Thon G, Bjerling P, Bünner CM, Verhein-Hansen J. Expression-state boundaries in the mating-type region of fission yeast. Genetics. 2002;**161**(2):611-622

[126] Sunder S, Greeson-Lott NT, Runge KW, Sanders SL. A new method to efficiently induce a site-specific double-strand break in the fission yeast *Schizosaccharomyces pombe*. Yeast. 2012;**29**(7):275-291

[127] Forsburg SL. Codon usage table for *Schizosaccharomyces pombe*. Yeast. 1994;**10**(8):1045-1047

[128] Wood V, Gwilliam R, Rajandream M-A, Lyne M, Lyne R, Stewart A, et al. The genome sequence of *Schizosaccharomyces pombe*. Nature. 2002;**415**(6874):871-880

[129] Castillo AG, Pidoux AL, Catania S, Durand-Dubief M, Choi ES, Hamilton G, et al. Telomeric repeats facilitate CENP-A(Cnp1) incorporation via telomere binding proteins. PLoS One. 2013;**8**(7):e69673

[130] Wellinger RJ, Zakian VA. Everything you ever wanted to know about *Saccharomyces cerevisiae* telomeres: Beginning to end. Genetics. 2012;**191**(4):1073-1105

[131] Jih G, Iglesias N, Currie MA, Bhanu NV, Paulo JA, Gygi SP, et al. Unique roles for histone H3K9me states in RNAi and heritable silencing of transcription. Nature. 2017;**547**(7664):463-467

[132] Gao Q, Reynolds GE, Wilcox A, Miller D, Cheung P, Artandi SE, et al. Telomerase-dependent and -independent chromosome healing in mouse embryonic stem cells. DNA Repair. 2008;**7**(8):1233-1249

[133] Muraki K, Han L, Miller D, Murnane JP. The role of ATM in the deficiency in nonhomologous end-joining near telomeres in a human cancer cell line. PLoS Genetics. 2013;**9**(3):e1003386

[134] Alcaraz Silva B, Jones TJ, Murnane JP. Differences in the recruitment of DNA repair proteins at subtelomeric and interstitial I-SceI endonuclease-induced DNA double-strand breaks. DNA Repair. 2017;**49**:1-8

[135] Doksani Y, de Lange T. Telomere-internal double-Strand breaks are repaired by homologous recombination and PARP1/Lig3-dependent end-joining. Cell Reports. 2016;**17**(6):1646-1656

[136] Mao P, Liu J, Zhang Z, Zhang H, Liu H, Gao S, et al. Homologous recombination-dependent repair of telomeric DSBs in proliferating human cells. Nature Communications. 2016;**7**(1):12154

[137] Liddiard K, Ruis B, Takasugi T, Harvey A, Ashelford KE, Hendrickson EA, et al. Sister chromatid telomere fusions, but not NHEJ-mediated inter-chromosomal telomere fusions, occur independently of DNA ligases 3 and 4. Genome Research. 2016;**26**(5):588-600

Chapter 4

Syndromes Associated with Telomere Shortening

Snehasish Nag

Abstract

We know that chromosomes are threadlike structures of nucleic acids and proteins, which are found in the nucleus of most living cells. They carry genetic information in the form of genes. Chromosomes are protected at their ends by a specialized structure called telomere. With each replicative cycle, the telomeres get shortened preventing uncontrolled replications. Telomeres perform several functions like protect the chromosome ends from sticking together, solve the end of replication problem, and limit the number of cell divisions. It is considered that telomeres are associated with cancer incidence and mortality. Telomere DNA has repetitive sequences (5′-TTAGGG-3′ in human), which is lengthened at the 3′ end by a special ribonucleoprotein enzyme called telomerase. Short telomeres are associated with early senescence, genomic instability, and apoptosis of cells. Short telomeres can result due to several factors including environmental factors, external factors like smoking, stress, as well as due to mutations in the components of telomere or telomerase. Short telomeres are associated with several disorders and diseases, such as dyskeratosis congenita, aplastic anemia, pulmonary fibrosis, and even cancer. Thus, it is important to understand how telomeres are associated with these diseases and what can be done to prevent such conditions.

Keywords: aplastic anemia, dyskeratosis congenita, idiopathic pulmonary fibrosis, telomere, telomerase

1. Introduction

Over the years, it has been observed that many degenerative disorders are associated with telomere dysfunction. Telomeres are present at the end of chromosomes. They protect the chromosome ends and critical genetic information in the chromosome from degradation by acting as caps from fusing with other chromosomes [1]. We know that the replication machinery cannot completely copy the chromosome ends, which is called end replication problem. As a result, the telomeres get shorter with each replicative cycle that leads to cell senescence [2]. Short telomeres are associated with genome instability. Telomere dysfunction caused by defects in telomerase proteins is associated with genomic instability that increases genetic mutations characterized by an increased incidence of cancer and also high sensitivity to genotoxic compounds. Short telomeres activate a p53-dependent checkpoint, which leads to senescence and apoptosis of the cells [3–6]. Telomere shortening can be caused by some external factors also such as smoking, stress, poor health such as obesity, inflammation [7]. Telomere shortening is also accelerated due to chemical and physical environmental agents. Reactive oxygen species can produce modified bases (mainly 8-oxoG) and single strand breaks in the genome. Oxidative damage can result from high incidence of guanine residues in telomeric DNA sequences [8]. Telomere shortening has been recognized as one of the important determinants behind senility and some diseases including—dyskeratosis congenita (DC), idiopathic pulmonary fibrosis (IPF) [9]. Telomere length is maintained by an enzyme called telomerase that adds telomeric repeats to the chromosome 3′-end using an RNA template. The enzyme is a ribonucleoprotein complex, which is inactive in somatic cells but active in stem cells and most cancer cells [10, 11]. Dysfunctional telomeres are recognized by many DNA damage response proteins leading to chromosome fusions, genome instability and altered gene expression patterns [12, 13]. Several cellular processes including apoptosis, aging, carcinogenesis, and chromosome instability are caused as consequences of loss of telomeres [14, 15].

2. Telomeres

Hermann J. Muller and Barbara McClintock in the 1930s described the telomere as a protective structure of DNA present at the end of the chromosome [16]. It protects the chromosome structure. The human telomeres have repetitive 5′-TTAGGG-3′ subunits, associated with a variety of telomere-associated proteins. The structure consists of a portion of the double-stranded DNA with an overhanging 3′ G-rich end (**Figure 1**) [1, 16].

Human somatic cells enter replicative senescence after a limited number of replications. This occurs due to the end replication problem leading to shortening of telomeres [17]. In absence of this structure, the replication cycle stops and the end-to-end fusion of chromosomes may occur [18–20]. Telomeres are bound by a specialized protein complex called shelterin [21–24]. Due to the end replication problem, the telomeres shorten with each cell cycle, and these short telomeres induce the DNA damage response and activate the p^{-53} dependent checkpoint, leading to apoptosis or senescence (**Figure 2A**) [21]. But in case of germ cells or in cancer cells, telomere maintenance is observed likely due to the expression or reactivation of telomerase, thus the replicative cycle of the cells continue.

The telomere shortening takes place as the eukaryotic DNA polymerases have no mechanism for synthesizing the final nucleotides present on the "lagging strand" of the double-stranded DNA. DNA polymerase synthesizes new DNA only from the 5′ → 3′ direction. The two strands of DNA are complementary, one strand is in 5′ → 3′ direction, while the other is in 3′ → 5′. DNA polymerase cannot synthesize DNA in the 3′ → 5′ direction. The process is compensated by the use of Okazaki fragments. Okazaki fragments are short pieces of DNA that are synthesized in the 5′ → 3′ direction from the 3′ → 5′ end as the replication fork moves. As RNA primer is required by DNA polymerase to synthesize new strand, each Okazaki fragment consists of an RNA primer followed by short DNA sequence. When the DNA polymerase reaches the chromosome end, the RNA primer is again placed, which

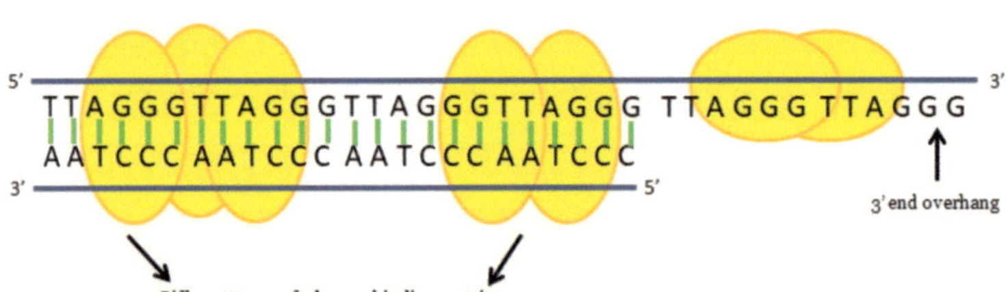

Figure 1.
3′ overhanging of telomere.

is inevitably removed. But as the primer is removed, the DNA polymerase cannot synthesize the remaining bases leading to telomere shortening with each replicative cycle (**Figure 2B**) [16, 25, 26].

In addition to that several external factors can also affect telomere length and maintenance. Factors such as smoking, alcohol consumption, chemical and environmental pollutants, radiation and many more can affect telomere length (**Figure 2C**).

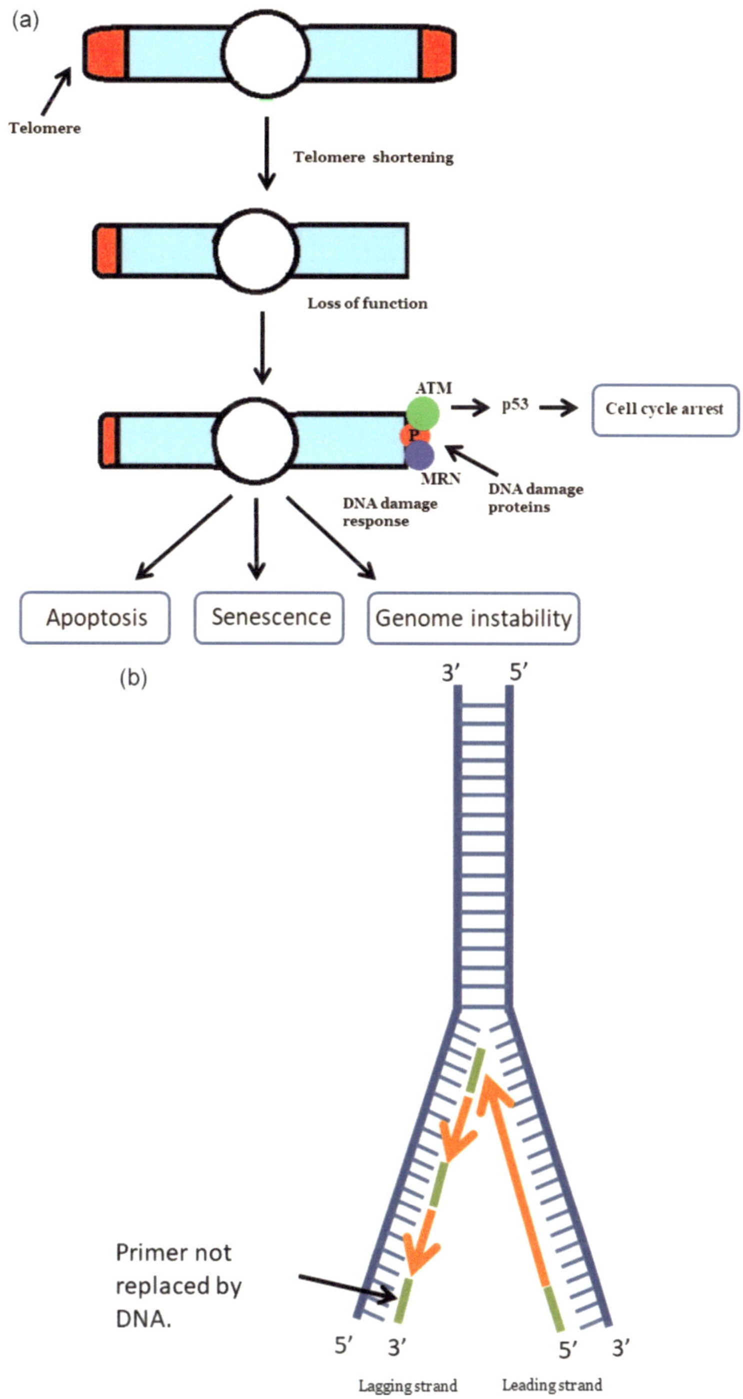

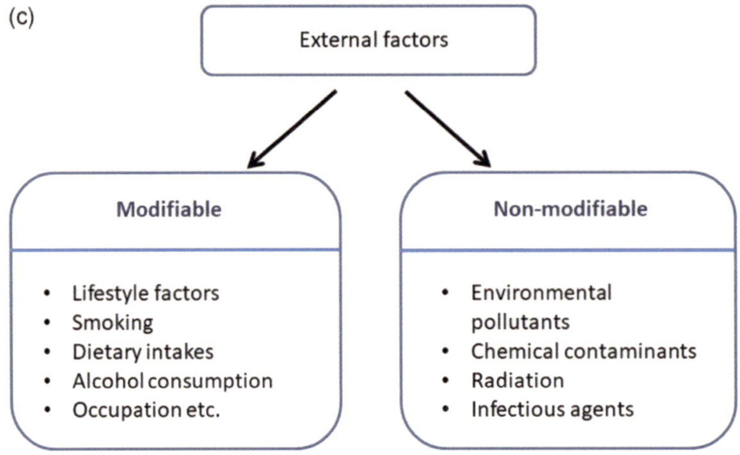

Figure 2.
(A) Telomere shortening leads to DNA damage response. The DNA damage responses include apoptosis, senescence of the cell or genomic instability that can lead to cancer. (B) "Lagging strand" end-replication problem. With each replication cycle the ends of the chromosome get shortened as the final RNA primer at the 3′-end cannot be replaced with DNA. (C) External factors associated with telomere shortening and maintenance.

3. Telomerase

The telomerase enzyme is a ribonucleoprotein containing both RNA and protein. It functions as a reverse transcriptase that positively regulates the telomere length [21, 27, 28]. The ribonucleoprotein has two essential components: telomerase reverse transcriptase (hTERT), the catalytic component, and telomerase RNA component (hTERC or hTR) which provides the template for telomere addition. Telomerase synthesizes new telomeres by solving the end-replication problem (**Figure 3**) [29].

Biogenesis of telomerase in somatic cells requires the assembly of hTERT and hTR into a stable complex that can function at telomeres. hTR (RNA component of telomerase) contains a box H/ACA motif which regulates RNA trafficking and

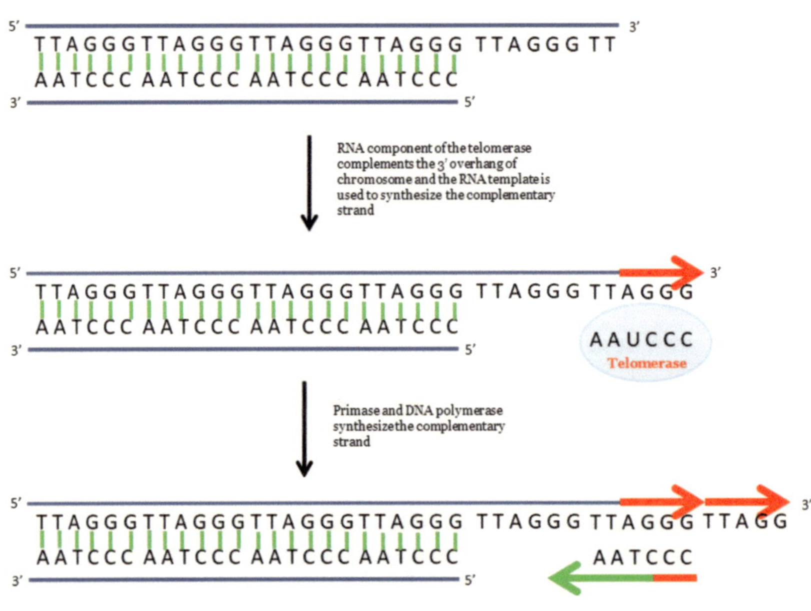

Figure 3.
An image showing how telomerase elongates telomere ends progressively.

stability. This H/ACA motif allows the hTR to associate with the dyskerin complex. This dyskerin complex is a four-protein core of dyskerin protein with another three nucleolar proteins—NOP10, NHP2, and GAR1 (**Figure 4**) [21, 30, 31]. Mutations in five out of six components that make up the telomerase ribonucleoprotein have been identified in humans causing telomere syndrome. These H/ACA RNAs can be divided into two groups. First, H/ACA small nucleolar RNAs (snoRNAs), that modifies ribosomal RNAs by accumulating in the nucleolus. Second, H/ACA small Cajal body-specific RNAs (scaRNAs) direct the modification of splicing RNAs by accumulating in Cajal bodies [32]. The difference in cellular trafficking between the two groups is attributable to the presence of another sequence motif, called Cajal body box or CAB box. They are the subnuclear sites of ribonucleoprotein assembly and modification [33]. The hTR has both H/ACA motif and also CAB box.

Shelterin component of telomerase regulates the synthesis of telomeres. It regulates the telomere length by forming t-loops whose formation is controlled by TRF2. TRF2 requires the help of other components such as TRF1 to function. Mutations in the shelterin components such as TRF2 and POT1 are found to be associated with short telomeres leading to such syndromes (**Figure 5**) [34].

A number of studies have revealed that in normal somatic cells the telomerase activity is almost absent. However a low level of telomerase activity has been found in mitotically active cells, including skin, lymphocytes, and endometrium. Telomerase enzyme is expressed in stem cells to maintain the telomere length all through their

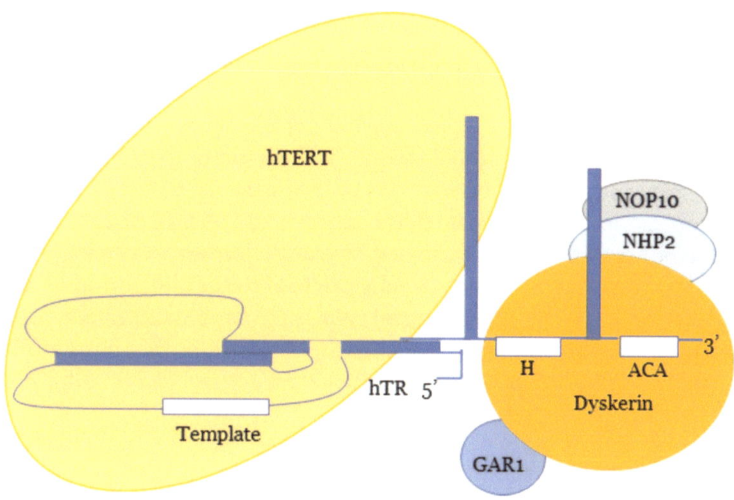

Figure 4.
The essential telomerase components.

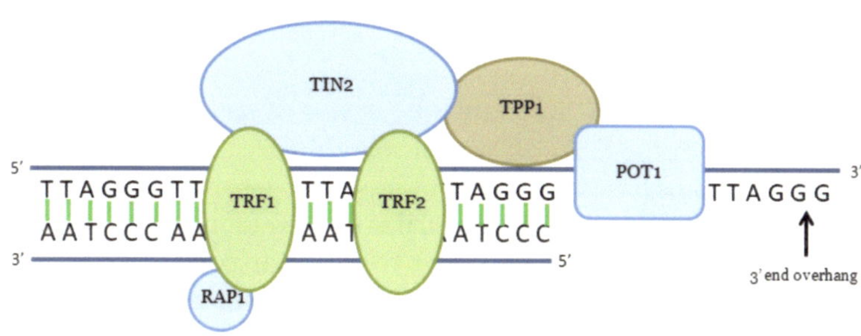

Figure 5.
The shelterin complex.

gene	Chromosome no.	function	Mode of Inheritance in dyskeratosis congenita
RTEL1	20	Helicase activity	Autosomal dominant and Autosomal recessive
TINF2	14	Forms part of shelterin complex	Autosomal dominant
hGAR1	4	Stability, maturation, localization	-
Dyskerin	Chromosome X	Stability, maturation, localization	X-linked
hNOP10	15	Telomere maintenance	Autosomal recessive
hNHP2	5	Stability, maturation, localization	Autosomal recessive
hStau	20	Accessibility to telomeres	-
hTERT	5	Synthesize protein component of telomerase	Autosomal dominant and Autosomal recessive
hTR	3	Synthesize RNA component of telomerase	Autosomal dominant

Table 1.
Human telomere shortening associated genes, their functions and mode of inheritance in dyskeratosis congenita [37].

life cycle. About 90% of the cancer cells have short telomeres with increased levels of telomerase activity [18]. For example, about 75% cases of oral carcinomas, 80% of lung cancers, 84% of prostate cancers, 85% of liver cancers, 93% of breast cancers, 94% of neuroblastomas, 95% of colorectal cancers, and 98% of bladder cancers have been found to be associated with increased levels of telomerase activity [35].

Telomerase transfection in normal cells can lead to the elongation of telomeres. For example, telomerase-negative normal cells, such as retinal pigment epithelial cells and foreskin fibroblasts, transfected with vectors encoding human hTERT show telomere elongation, but telomerase-negative control cells exhibit both telomere shortening and senescence [36].

Furthermore, mutations in the telomerase and telomere components lead to the syndromes of telomere shortening (**Table 1**).

4. Syndromes associated with short telomere

4.1 Dyskeratosis congenita (DC)

Dyskeratosis congenita (DC) is a rare progressive congenital disorder having a highly variable phenotype [38]. DC is a rare syndrome of premature aging. The term coined by clinicians based on a triad of mucocutaneous features that they found in male children. These are—leukoplakia of the oral mucosa, skin hyperpigmentation, and dystrophy of nails [39]. This triad was associated with premature mortality of children due to bone marrow failure in aplastic anemia. DC mainly affects the skin. But in nearly 80% of the cases, bone marrow failure also occurs. DC is also characterized by the predisposition of cancer. In serious forms, the life span can be significantly shortened.

4.1.1 Genetics of the syndrome

In 1998, the gene encoding dyskerin, DKC1 was discovered. It was identified in X-linked families with the help of linkage and positional cloning. Dyskerin is a putative box H/ACA telomerase RNA binding protein [40]. The protein links with the telomerase RNA structure. The hTR has a box H/ACA motif and the X-linked DC patients have low levels of telomerase RNA component resulting in short telomeres. It is supported by the fact that mutations in the DKC1 gene disrupt the maturation and stability of hTR. Mutations in the dyskerin complex, NOP1O and NHP2 have also been identified in DC families [41, 42].

The best characterized form of dyskeratosis congenita is a result of one or more mutations in the gene DKC1 present on the long arm of X chromosome. This result in the X-linked recessive form of the disease also called Zinsser-Cole-Engman syndrome wherein the major protein affected is dyskerin [40]. Within the vertebrates, dyskerin is a key component of the telomerase RNA component (hTR) in the form of the H/ACA motif. This X-linked variety, like the NOP10 and NHP2 mutations, demonstrates shortened telomeres as a result of lower hTR concentrations [43, 44].

Recently, heterozygous mutations in the shelterin component TINF2 were identified in several cases of DC. Mutations in the TINF2 results in severe manifestations and usually present in children [34, 45]. Different organs show different types of defects in DC patients (**Table 2**).Many of these cells express telomerase, an enzyme that maintains telomeres.

Mutations in DKC1 can lead to significant declines in hTR levels, i.e. one fifth of the wild-type [46]. This is consistent with the fact that mutations in the DKC1

Organ System	Telomerase expressing cells	Defects in Dyskeratosis congenita
Hair	Hair follicle	Alopecia
Oral cavity	Squamous epithelium	Leukoplakia
Skin	Basal layer of epidermis	Hyperpigmentation, Nail dystrophy
Lungs	Type 2 alveolar epithelial cells	Fibrosis
Liver	Distributed hepatocytes	Cirrhosis
Intestine	Intestinal crypts	Disorders of gut
Testes	Spermatogonia	Hypogonadism
Bone marrow	Progenitor stem cells	Blood cell production failure

Table 2.
Defects in DC patients are most often seen in tissues in which cells divide rapidly, and often, many of these cells express telomerase, an enzyme that maintains telomeres.

lead to accelerated phenotypes because of a loss of greater than half of the available telomerase. Mutations in the shelterin component TINF2 also lead to severe disease. This suggests that telomere defects are alone sufficient to cause dyskeratosis congenita (DC) [40, 42, 43].

Due to aplastic anemia when DC patients undergo bone marrow transplant, they frequently suffer with morbidity and mortality from pulmonary fibrosis and liver failure. This happens even when the patients seem to have intact function in these organs during the time of transplant [47, 48]. This happens due to the limited length of the telomeres in the patient's lung and liver, and also the poor capacity of DNA damage repair after chemotherapy and radiation.

Nonmyeloablative bone marrow transplant should be considered in aplastic anemia, where there is mutation in the telomere or telomerase components [40].

4.1.2 DC patients are cancer prone

As many as 10% of the DC patients die due to the cancer diagnosis. DC is thought to be a cancer-prone disorder because of the underlying pathology of abnormal telomere maintenance. The link between DC and cancer is very interesting, because DC is associated with defects in telomere biology. Patients with DC have very short telomeres. Mutations have been identified in telomere biology genes. The United Kingdom Dyskeratosis Congenita Registry (DCR) data indicated that the crude rate of malignancy among approximately 300 patients was 10%. DC patients are at increased risk of myelodysplasia and acute leukemia [49]. Since aplastic anemia itself has an associated increased risk for transformation to acute myeloid leukemia, it is unclear whether DC patients with aplastic anemia have an added predisposition. DC patients also have increased incidence of squamous cell cancers of the skin and head and neck. In DC patients, these cancers are diagnosed at as early as the 2nd decade of the life. DC patients with cancer have a mean age at cancer diagnosis of 29 and a cumulative incidence of ~40% by the age of 50.

4.1.3 Predisposition to cancer

Susceptibility to cancer seems counterintuitive due to the fact that in many known cancers reactivation of telomerase is actually a required step for malignancy to evolve however short telomeres do contribute to genome instability. In a disease like DC where telomerase is affected, it does not seem that cancer would be a complication to result. But it is discussed that with critically short or absent telomeres, chromosomes will likely be attached together at their ends through the non-homologous end joining pathway (NHEJ). If this occurrence is common enough, then malignancy even without functional telomerase seems probable.

4.1.4 Haploinsufficiency of telomerase

Families with autosomal dominant dyskeratosis congenita show anticipation and have mutations in the telomerase RNA gene. A null mutation in motif D of the hTERT domain is associated with this phenotype. This mutation leads to haploinsufficiency of telomerase, and telomere shortening occurs despite the presence of telomerase (**Figure 6**) [50].

This finding shows the importance of telomere maintenance and telomerase dosage for maintaining tissue proliferative capacity. It has also relevance for understanding mechanisms of age-related changes. Telomere length limits the number of replication cycle of primary fibroblasts and has been associated with

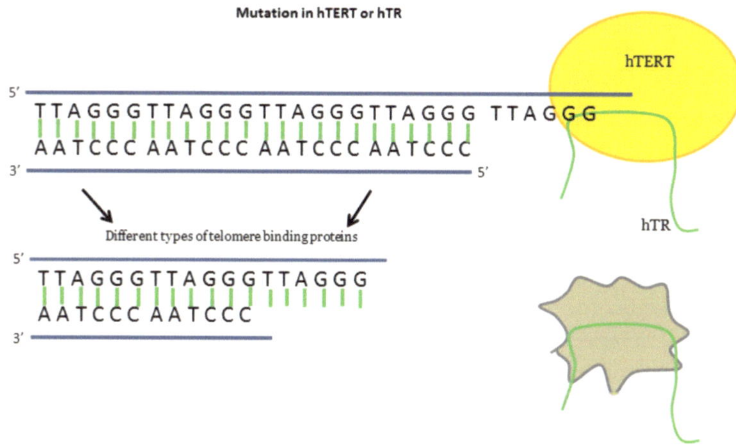

Figure 6.
Telomere shortening despite the presence of telomerase. Mutations in the hTERT or hTR components of telomerase prevent them from extending the telomere length.

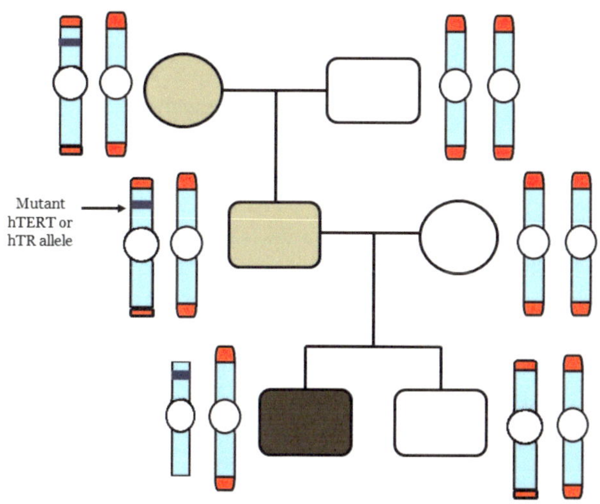

Figure 7.
The figure depicts autosomal dominance mode of inheritance. The dark blue region represents mutant hTERT or hTR allele. Darker shades of black represent progressive telomere shortening leading to anticipation of phenotypes that is the age of onset becomes earlier with each generation.

cellular aging [50]. Short telomeres activate DNA damage response, which leads to apoptosis. It is the shortest telomere and not the average telomere length within a cell that is responsible for mediating the response that leads to cell death [51]. Mutations in the hTERT component can result in a complex phenotype of stem cell failure. This phenotype shows anticipation; it presents earlier and more severely with successive generations. The anticipation is due to haploinsufficiency of telomerase that results in progressive shortening of telomeres (**Figure 7**) [52]. The hTERT mutation results in haploinsufficiency of telomerase, which leads to shortening of telomeres across generations [49]. The number of these short telomeres is correlated with the severity of phenotypes expressed. The earlier onset of such phenotypes in later generations implicates that in bone marrow and other solid tissues the telomere length is short and in limiting proliferative capacity. This pattern of anticipation suggests that like aplastic anemia, this disorder might also affect the stem cells within the lung.

4.2 Aplastic anemia due to telomere shortening

Aplastic anemia arises when the body's bone marrow does not make enough new blood cells. It can develop at any age. A number of diseases, conditions and factors can damage the blood-making stem cells in bone marrow and bring about aplastic anemia.

Aplastic anemia patients with shorter chromosome tips, or telomeres, have a lower survival rate and are much more likely to relapse after treatment than those with longer telomeres. Studies identified germline mutations in the hTR and hTERT components of the telomerase in ~3% of the adults with so-called aplastic anemia [52, 53]. In recent years, scientists have found that some patients suffering with severe aplastic anemia have extremely short telomeres in their blood cells. Telomeres are also known as molecular caps that protect the chromosomes ends from erosion. With each cell division they naturally become shorter, but telomeres can be rebuilt by enzymes. Telomere length is affected by genetic factors and environmental stressors. Patients with short telomeres suffer from morbidity and mortality even after the bone marrow transplant for the aplastic anemia [47, 54]. As these short telomeres lead to organ failures.

Patients with the short telomeres are also at greater risk for a conversion to bone marrow cancer (24%).

4.3 Idiopathic pulmonary fibrosis due to telomere shortening

Idiopathic pulmonary fibrosis has a predictable and progressive clinical course that ultimately leads to respiratory failure [55]. Although both genetic and environmental factors have been implicated, the cause of idiopathic pulmonary fibrosis (IPF) is unknown. IPF is the most common manifestation out of the other telomere-mediated disorders [56]. Germ line mutations in the telomerase hTERT and hTR component genes are the reason behind up to one-sixth of pulmonary fibrosis families [57]. The presence of telomerase mutations is significant. As extra-pulmonary complications, affected individuals can suffer from bone marrow failure and cryptogenic liver cirrhosis due to telomere shortening. Evidence suggests that IPF results from autosomal dominant telomere syndromes. Here with successive generations, the condition evolves from pulmonary fibrosis to a disorder of bone marrow failure. It is perhaps the most devastating of the idiopathic disorders in medicine.

IPF is an age-related disease. From the time of diagnosis, IPF patients live on average 3 years. Several clinical factors are known which are associated with the IPF. Age is the biggest with the great majority are diagnosed after the age of 60. It is also most frequently in males with a nearly 2:1 ratio [58].

4.3.1 IPF in dyskeratosis congenita patients

DC represents a more severe presentation of a spectrum of telomere syndromes where IPF represents an attenuated form [40]. Pulmonary fibrosis in case of bone-marrow failure can be precipitated by pulmonary toxic drugs during of bone marrow transplant. For example, fatal pulmonary fibrosis in DC patients is caused by the alkylating agent busulfan used in myeloablative conditioning regimens [59]. Even without precipitating toxins, pulmonary fibrosis is a significant and under-estimated complication of DC. In some DC patients, pulmonary fibrosis is the major cause of premature mortality in the absence of bone marrow failure [60].

4.3.2 IPF is the most frequent manifestation of telomere-associated disease

In most cases, IPF is associated with telomere maintenance. Mutations in hTERT and hTR are the risk factors in 8–15% of familial cases of IPF [57, 61]. In about 3% of sporadic IPF cases, mutations in the telomerase genes are also found [53]. Here hTERT mutations frequency is higher than hTR mutations, but the mutant genes cannot be identified based on only clinical features [62]. Short telomeres are sufficient to cause the common form of IPF [63].

The hTERT and hTR mutations result in short telomeres because of the loss of functions and the haploinsufficiency [56]. As compared to DC and aplastic anemia, the prevalence of IPF is more common, lung disease is the most common manifestation of telomere-mediated disorders [64, 65]. Thus, although DC is specific for identifying individuals with telomere-mediated disease, it only can identify only a small subset, i.e. nearly 5% of all cases.

4.3.3 IPF patients with short telomeres without any mutations in telomerase

Although telomerase mutations are found in one-sixth of the families with IPF, short telomeres are found in other IPF patients without any mutations in the telomerase genes [58]. Significantly shorter telomeres are seen in case of sporadic IPF cases (those who report no family history) [61, 65]. Telomere shortening can be found in immune cells such as lymphocytes, granulocytes and also alveolar epithelial cells, which implicate global telomere defect in such individuals. The observation suggests that individuals with shortest telomeres are more likely to develop IPF than normal individuals in the population [66]. These patients with short telomeres may be a risk factor for disease outside the lung. A subset of sporadic IPF that lack an apparent telomerase mutation also develops cryptogenic liver cirrhosis [57]. There is also relation between IPF and incidence of diabetes. IPF patients have about 3-fold increased incidence of diabetes compared to the age-matched controls [67]. In case of telomerase deficient mice, short telomeres cause defects in insulin secretion resulting in glucose intolerance. Therefore alongside IPF, short telomeres can be a risk factor for diabetes development [68]. Thus in sporadic IPF cases, the defect in telomere length may cause telomere-associated diseases outside of lung.

4.3.4 IPF patients with extra-pulmonary disorders

IPF patients and their relatives who carry telomerase mutations can develop telomere-mediated diseases, which are extra-pulmonary [57]. These are, bone-marrow failure including macrocytosis of red blood cells, single lineage cytopenias, aplastic anemia, myelodysplastic syndromes, and acute myeloid leukemia [69, 70, 71]. In case of patients without DC, IPF and bone marrow failure are not considered as related conditions. But occurrence of these two together allows clinical identification of families carrying telomerase mutations. A recent finding suggests that germ line defects in telomerase of a single family are associated with the occurrence of these two disorders together [62]. When present in successive generations, both the IPF and bone marrow failure syndrome together predicted the presence of an hTERT or hTR gene mutation in 10 out of 10 families (100%).

Other than the bone marrow failure, IPF patients with telomerase mutations may also develop other complications of telomere-mediated disease like liver cirrhosis [50]. So, the IPF affected individuals are at a higher risk of developing extra-pulmonary diseases.

4.4 Role of telomeres and telomerase in cancer

Short telomeres due to mutations in telomerase have been proposed to be associated with cancer. The concept seems counterintuitive as we know that telomerase activation is a required step for malignancy to occur in nearly 85% of the cases as it allows unlimited cell cycle without senescence. How telomerase reactivation occurs in case of cancer is not clear till date. Studies suggest mutations in two key positions of hTERT promoter region (C250T and C228T) cause enhanced expression of hTERT leading to enhanced telomerase activation (**Table 3**) [72, 73]. But this information needs to be investigated properly. Short telomeres can lead to genomic instability and also cancer via non-homologous end joining (NHEJ) of chromosomes. Mutations in the hTR or hTERT components of telomerase are associated with abnormally short telomeres leading to cancer. Mutations in several components of telomerase such as DKC1, NOP10, NHP2, GAR1 or shelterin components such as TRF1, TRF2, POT1 can lead to short telomeres [30]. Absence or very short telomeres allow non-homologous chromosomes to join head to head. Syndromes associated with short telomeres such as dyskeratosis congenita, aplastic anemia are associated with cancer. It has been found that DC patients are cancer prone. They have increased risk of acute leukemia and myelodysplasia [49]. Aplastic anemia is also associated with acute myeloid leukemia. Patients also have increased risk of squamous cell cancers. Overexpression of TERT is also associated with increased cell proliferation in epidermal tumors and mammary carcinomas in mice [74].

Thus genomic instability due to loss of telomeres and overexpression of telomerase probably play major roles in such cancer development.

	Types of cancer	Occurrence (~)
Mutations in the hTERT promoter region	melanomas	70 %
	glioblastomas	80-90%
	hepatocellular carcinomas	60%
	bladder cancers	60%
	basal cell carcinomas	70%
	cutaneous squamous cell carcinomas	50%
	thyroid cancers	30%
	oligodendrogliomas	72%

Table 3.
Types of cancers associated with mutations in the telomerase hTERT promoter region.

5. Conclusion

Cellular aging eventually leads to cell death. It is the progressive decline of cells in resisting stress and other cellular damages. This leads to gradual loss of cellular functions resulting in cell death. Telomere shortening is a major factor that is related with cellular aging. With age, the telomere length declines due to end replication problem, leading to cell senescence. It poses a barrier to the tumor growth but also results in the loss of cells with aging. When the caps of the chromosomes which are telomeres become critically short, it prevents cell cycle to continue leading to either cell senescence or apoptosis. This cell cycle arrest occurs due to DNA damage proteins such as ATM, which become activated when telomere becomes critically shortened leading to activation of p53 dependent checkpoint. Mutations in the telomere or the telomerase components such as hTR or hTERT result in a broad spectrum of diseases present in children and adults. The onset and severity of these diseases are determined by the extent of telomere shortening. Usually the onset of cancer is associated with the activity of the telomerase holoenzyme, but with reduced telomeres due to affected telomerase, the chromosomes may join by non-homologous end joining (NHEJ) and can lead to malignancy. This study shows that syndromes such as dyskeratosis congenita (DC), idiopathic pulmonary fibrosis (IPF), and aplastic anemia are caused by the telomere shortening. IPF syndrome is the most common manifestation of the telomere shortening. Thus this provides evidence that short telomeres are sufficient to cause common, age-related diseases. Treatment for these diseases involves organ transplantation such as liver, lung, bone marrow. Although this organ transplantation provides improved physical condition for patients, it does not address the actual cause, which is short telomeres. In recent times, telomerase activators such as TA-65 has gained commercial interest. It is also reported that sex hormones activate TERT transcription.

The understanding of the role of telomere and telomerase in aging and some diseases can open new possibilities in understanding the genetic factors that play important role in the origin and augmentation of several other diseases.

Acknowledgements

SN is thankful to University Grants Commission, New Delhi, India. The author is thankful to Dr. Rakesh Kundu for his technical assistance and constant encouragement.

Conflict of interest

The author declares no conflict of interest.

Author details

Snehasish Nag
Visva-bharati University, Santiniketan, West Bengal, India

*Address all correspondence to: snehasishn@gmail.com

References

[1] Armanios M et al. The telomere syndromes. Nature Reviews Genetics. 2012;**13**(10):693-704

[2] Watson JD et al. Origin of concatemeric T7DNA. Nature—New Biology. 1972;**239**:197-201

[3] Artandi SE et al. Telomere dysfunction promotes non-reciprocal translocations and epithelial cancers in mice. Nature. 2000;**406**:641-645

[4] Feldser DM et al. Short telomeres limit tumor progression in vivo by inducing senescence. Cancer Cell. 2007;**11**:461-469

[5] Harley CB et al. Telomeres shorten during ageing of human fibroblasts. Nature. 1990;**345**:458-460

[6] Lee HW et al. Essential role of mouse telomerase in highly proliferative organs. Nature. 1998;**392**:569-574

[7] Shay JW. Role of telomeres and telomerase in aging and cancer. Cancer Discovery. 2016;**6**(6):584-593

[8] Coluzzi E et al. Oxidative stress induces persistent telomeric DNA damage responsible for nuclear morphology change in mammalian cells. PLoS One. 2014;**9**(10):110963

[9] Frescas D et al. A TIN2 dyskeratosis congenita mutation causes telomerase-independent telomere shortening in mice. Genes and Development. 2014;**28**(2):153-166

[10] Greider CW et al. Identification of a specific telomere terminal transferase activity in *Tetrahymena* extracts. Cell. 1985;**43**:405-413

[11] Shay JW et al. A survey of telomerase activity in human cancer. European Journal of Cancer. 1997;**33**:787-791

[12] de Lange T. Protection of mammalian telomeres. Oncogene. 2002;**21**:532-540

[13] Blackburn EH et al. Molecular manifestations and molecular determinants of telomere capping. Cold Spring Harbor Symposia on Quantitative Biology. 2000;**65**:253-263

[14] Harley CB. Telomere loss: Mitotic clock or genetic time bomb? Mutation Research. 1991;**256**:271-282

[15] Bailey SM. Telomeres, chromosome instability and cancer. Nucleic Acids Research. 2006;**34**:2408-2417

[16] Wai LK. Telomeres, telomerase, and tumorigenesis—A review. Medscape General Medicine. 2004;**6**:19

[17] Olovnikov AM. A theory of marginotomy: The incomplete copying of template margin in enzymic synthesis of polynucleotides and biological significance of the phenomenon. Journal of Theoretical Biology. 1973;**41**(1):181-190

[18] Shay JW et al. Telomeres and telomerase in normal and cancer stem cells. FEBS Letters. 2010;**584**:3819-3825

[19] Shay JW, Wright WE. Historical claims and current interpretations of replicative aging. Nature Biotechnology. 2002;**20**:682-688

[20] Greider CW et al. Telomeres, telomerase and cancer. Scientific American. 1996;**274**(2):92-97

[21] Artandi SE. Telomeres and telomerase in cancer. Carcinogenesis. 2010;**31**:9-18

[22] de Lange T. Shelterin: The protein complex that shapes and safeguards human telomeres. Genes & Development. 2005;**19**:2100-2110

[23] Baumann P et al. Pot1, the putative telomere end-binding protein in fission yeast and humans. Science. 2001;**292**:1171-1175

[24] Liu D et al. PTOP interacts with POT1 and regulates its localization to telomeres. Nature Cell Biology. 2004;**6**:673-680

[25] Karp G. DNA replication and repair. In: Cell and Molecular Biology. 2nd ed. New York, NY: John Wiley & Sons, Inc; 1999. pp. 575-607

[26] Reddel RR. The role of senescence and immortalization in carcinogenesis. Carcinogenesis. 2000;**21**:477-484

[27] Meyerson M et al. hEST2, the putative human telomerase catalytic subunit gene, is up-regulated in tumor cells and during immortalization. Cell. 1997;**90**:785-795

[28] Weinrich SL et al. Reconstitution of human telomerase with the template RNA component hTR and the catalytic protein subunit hTRT. Nature Genetics. 1997;**17**:498-502

[29] Blasco MA. Mammalian telomeres and telomerase: Why they matter for cancer and aging. European Journal of Cell Biology. 2003;**82**:441-446

[30] Mitchell JR et al. A box H/ACA small nucleolar RNA-like domain at the human telomerase RNA 3′ end. Molecular and Cellular Biology. 1999;**19**:567-576

[31] Chen JL et al. Secondary structure of vertebrate telomerase RNA. Cell. 2000;**100**:503-514

[32] Matera AG et al. Non-coding RNAs: Lessons from the small nuclear and small nucleolar RNAs. Nature Reviews Molecular Cell Biology. 2007;**8**:209-220

[33] Cioce M et al. Cajal bodies: A long history of discovery. Annual Reveiw of Cell and Development Biology. 2005;**21**:105-131

[34] Walne AJ. TINF2 mutations result in very short telomeres: Analysis of a large cohort of patients with dyskeratosis congenita and related bone marrow failure syndromes. Blood. 2008;**112**:3594-3600

[35] Belair CD et al. Telomerase activity: A biomarker of cell proliferation, not malignant transformation. Proceedings of the National Academy of Sciences of the United States of America. 1997;**94**(25):13677-13682

[36] Kuznetsova AV et al. Cell models to study regulation of cell transformation in pathologies of retinal pigment epithelium. Journal of Ophthalmology. 2014;**2014**:801787

[37] Mason PJ. The genetics of dyskeratosis congenita. Cancer Genetics. 2011;**204**(12):635-645

[38] James W et al. Andrews' Diseases of the Skin: Clinical Dermatology. 7th ed. Philadelphia: Saunders Elsevier; 2006

[39] Dokal I et al. Dyskeratosis congenita: Its link to telomerase and aplastic anaemia. Blood Reviews. 2003;**17**:217-225

[40] Armanios M. Syndromes of telomere shortening. Annual Review of Genomics and Human Genetics. 2009;**10**:45-61

[41] Vulliamy T et al. Association between aplastic anaemia and mutatons in telomerase RNA. The Lancet. 2002;**359**:2168-2170

[42] Wong JM et al. Telomerase RNA level limits telomere maintenance in X-linked dyskeratosis congenita. Genes & Development. 2006;**20**:2848-2858

[43] Mitchell JR et al. A telomerase component is defective in the human

disease dyskeratosis congenita. Nature. 1999;**402**:551-555

[44] Ball SE et al. Progressive telomere shortening in aplastic anemia. Blood. 1998;**91**:3582-3592

[45] Savage SA et al. TINF2, a component of the shelterin telomere protection complex, is mutated in dyskeratosis congenita. American Journal of Human Genetics. 2008;**82**:501-509

[46] Ruggero D et al. Dyskeratosis congenita and cancer in mice deficient in ribosomal RNA modification. Science. 2003;**299**:259-262

[47] de la Fuente J et al. Dyskeratosis congenital: Advances in the understanding of the telomerase defect and the role of stem cell transplantation. Pediatric Transplantation. 2007;**11**:584-594

[48] Rocha V et al. Unusual complications after bone marrow transplantation for dykeratosis congenital. British Journal of Haematology. 1998;**103**:243-248

[49] Alter BP. Diagnosis, genetics, and management of inherited bone marrow failure syndromes. Hematology. American Society of Hematology. Education Program. 2007;**2007**:29-39

[50] Armanios M et al. Haploinsufficiency of telomerase reverse transcriptase leads to anticipation in autosomal dominant dyskeratosis congenita. Proceedings of the National Academy of Sciences of the United States of America. 2005;**102**(44):15960-15964

[51] Hemann MT et al. Telomere dysfunction triggers developmentally regulated germ cell apoptosis. Molecular Biology of the Cell. 2001;**12**:2023-2030

[52] Yamaguchi H et al. Mutations inTERT, the gene for telomerase reverse transcriptase, in a plastic anemia. The New England Journal of Medicine. 2005;**352**(14):1413-1424

[53] Yamaguchi H et al. Mutations of the human telomerase RNA gene (TERC) in aplastic anemia and myelodysplastic syndrome. Blood. 2003;**102**:916-918

[54] Yabe M et al. Fatal interstitial pulmonary disease in a patient with dyskeratosis congenital after allogeneic bone marrow transplantation. Bone Marrow Transplantation. 1997;**19**:389-392

[55] Armanios MY et al. Telomerase mutations in families with idiopathic pulmonary fibrosis. The New England Journal of Medicine. 2007;**356**(13):1317-1326

[56] Alder J et al. Short telomeres are a risk factor for idiopathic pulmonary fibrosis. Proceedings of the National Academy of Sciences of the United States of America. 2008;**105**:13051-13056

[57] Loyd JE. Pulmonary fibrosis in families. American Journal of Respiratory Cell and Molecular Biology. 2003;**29**:47-50

[58] Armanios M. Telomerase and idiopathic pulmonary fibrosis. Mutation Research. 2012;**730**:52-58

[59] Gungor T et al. Nonmyeloablative allogeneic hematopoietic stem cell transplantation for treatment of dyskeratosis congenita. Bone Marrow Transplantation. 2003;**31**:407-410

[60] Parry EM et al. Decreased dyskerin levels as mechanism of telomere shortening in Xlinked dyskeratosis congenital. Journal of Medical Genetics. 2011;**48**:327-333

[61] Tsakiri KD et al. Adultonset pulmonary fibrosis caused by mutations in telomerase. Proceedings of the National Academy of Sciences of the United States of America. 2007;**104**:7552-7557

[62] Parry EM et al. Syndrome complex of bone marrow failure and pulmonary fibrosis predicts germline defects in telomerase. Blood. 2011;**117**:5607-5611

[63] Raghu G et al. Incidence and prevalence of idiopathic pulmonary fibrosis. American Journal of Respiratory and Critical Care Medicine. 2006;**174**:810-816

[64] Kirwan M et al. Dyskeratosis congenita: A genetic disorder of many faces. Clinical Genetics. 2008;**73**:103-112

[65] Cronkhite JT et al. Telomere shortening in familial and sporadic pulmonary fibrosis. American Journal of Respiratory and Critical Care Medicine. 2008;**178**:729-737

[66] Aviv A. Genetics of leukocyte telomere length and its role in atherosclerosis. Mutation Research. 2012;**730**(1-2):68-74

[67] Gribbin J et al. Role of diabetes mellitus and gastrooesophageal reflux in the aetiology of idiopathic pulmonary fibrosis. Respiratory Medicine. 2009;**103**:927-931

[68] Guo N et al. Short telomeres compromise betacell signaling and survival. PLoS One. 2011;**6**:17858

[69] Diaz de Leon A et al. Telomere lengths, pulmonary fibrosis and telomerase (TERT) mutations. PLoS One. 2010;**5**:10680

[70] Du HY et al. TERC and TERT gene mutations in patients with bone narrow failure and the significance of telomere length measurement. Blood. 2009;**113**:309-316

[71] Kirwan M et al. Defining the pathogenic role of telomerase mutations in myelodysplastic syndrome and acute myeloid leukemia. Human Mutation. 2009;**30**(11):1567-1573

[72] Akincilar SC et al. Reactivation of telomerase in cancer. Cellular and Molecular Life Sciences. 2016;**73**(8):1659-1670

[73] Huang FW et al. TERT promoter mutations and monoallelic activation of TERT in cancer. Oncogene. 2015;**4**:e176

[74] Gonzalez-Suarez E et al. Increased epidermal tumors and increased skin wound healing in transgenic mice overexpressing the catalytic subunit of telomerase, mTERT, in basal keratinocytes. The EMBO Journal. 2001;**20**:2619-2630

Chapter 5

Telomeres and Telomerase Activity in the Human Placenta

Marie Jirkovská, Marie Korabečná and Soňa Laššáková

Abstract

Placenta is a transient organ ensuring the intrauterine development of the individual. To meet fetal requirements, rapid and continuous cell proliferation enlarges the areas of tissues maintaining maternofetal transport. The cell division in placenta is accompanied with shortening of telomeres leading to cell senescence. Telomerase activity, on the other hand, ensures replication of telomeres and allows the organ to serve till the end of pregnancy. This balanced process may be negatively influenced by unfavorable circumstances. Here, we summarize available data on telomere length as well as telomerase activity in placentas from normal and complicated pregnancies; attention is also paid to the comparison of methods used in relevant studies.

Keywords: pathology, placenta, pregnancy, telomerase, telomere

1. Introduction

The core of cell proliferation is the division of cells and replication of chromosomes. Among other factors, it is also regulated by the length of telomeres since short telomeres will either recruit telomerase, or, in the absence of telomerase, induce senescence, apoptosis, or genome instability, or activate a DNA damage response (e.g., telomere recombination). The main function of telomeres is the protection of chromosomal integrity during DNA replication; moreover, they themselves are protected by a shelterin protein complex. Telomeres stabilize the ends of linear chromosomes and prevent the ends from being recognized as a double strand break. In human cells, telomeres contain hexameric tandem repeats, 5′TTAGGG 3′, of DNA sequence. To maintain the proliferative ability of cells, the elongation of telomeres is executed by adding telomeric DNA repeats to the 3′chromosomal ends by telomerase. In the absence of telomerase, the telomeres shorten in every cell division. Telomerase as an RNA-dependent DNA polymerase repairs the sequences of telomeres after each cell division; but in humans, this enzyme is active in stem cells, germ cells [1–3], and cancer cells only.

During development of an individual, mature oocytes and cleavage stage embryos display low or absent telomerase activity, whereas in the blastocyst stage, its activity is high again. As in cells during early cleavage, the telomeres become remarkably longer, and an alternate lengthening of telomeres may play a role in their elongation. Processes of telomeric DNA recombination between homologous sister chromatids take place in the cleavage stage, and the length of telomeres is then maintained from the blastocyst stage onward by telomerase [4]. Telomeric DNA recombination between telomeres on separate chromosomes such as gene conversion and the elongation of telomeres by DNA polymerase activity by mechanisms like break-induced replication can lengthen telomeres independently of telomerase activity [5].

The mechanism of DNA replication results in progressive shortening of the ends of linear DNA molecule. That shortening limits the life span of individual cells and it is referred to as replication senescence [6]. Recently, it is well known that telomere attrition is observed during normal cellular aging, but telomere dysfunction may also contribute to the onset and progression of age-related diseases like atherosclerosis and myocardial infarction [7]. Telomere attrition is regarded as one of the so-called hallmarks of aging as proposed in [8].

Many tissues and organs contain cycling undifferentiated stem cells that provide cells for their renewal. The frequency of their mitotic division is different among tissues; examples of the most active are epidermal cells and cells of the bone marrow. Those cells are characterized by high levels of telomerase expression and disorders of telomere homeostasis cause, or at least take part in, the pathogenesis of serious inherited diseases, for example, dyskeratosis congenita [9], aplastic anemia, other bone marrow syndromes [10–12] and/or idiopathic pulmonary fibrosis [13]. Moreover, increased incidence of diabetes mellitus was identified in patients suffering from those pathologies [14]. Inherited bone marrow failure syndromes threaten also the prenatal development due to fetal malformations and intrauterine growth retardation accompanied with an abnormally small placenta [15].

The correct function of telomerase-telomere complex depends on both genetic predispositions and external factors (age, reactive oxygen species, and exogenous genotoxic factors). Mutations in the telomerase holoenzyme in either of the two genes, *TERT* encoding the reverse transcriptase, or *TERC* encoding the RNA template for the synthesis of telomeres by telomerase, can cause remarkable telomere attrition even in hemizygous individuals [16] and may take part in hereditary conditioned telomere disorders [17–20].

2. Placenta in pregnancy

The prenatal development of an individual is conditioned by placenta, the transient organ that functions exclusively for the time of pregnancy. In order to meet fetal requirements, the placenta holds the functions of still undeveloped fetal organs, for example, lung and kidney. It is the site of transport of oxygen, ions, nutrients, and maternal immunoglobulins from mother to fetus, and carbon dioxide and wastes from fetus to mother. The placenta also maintains pregnancy by production of steroid and protein hormones and other factors. Due to its position between maternal and fetal bloodstreams, the placenta acts as a barrier against infectious agents and regulates the maternal immune tolerance, gas exchange, and fetal nutrition. On the other hand, its structure and function are negatively impacted by maternal and fetal metabolic disturbances in pathological pregnancies.

Placental tissues originate in extraembryonic structures, that is in trophoblast, the outer layer of the blastocyst, which invades maternal tissues and gives rise to the cytotrophoblast and syncytiotrophoblast, and in extraembryonic mesoderm that is requisite for the formation of placental vasculature and supporting connective tissue.

The progress of fetal growth and maturation of fetal organs is essentially accompanied by the growth of placental size. It is performed by continuous cell proliferation till the term of gestation [21] and balanced with differentiation and apoptosis in all tissue compartments. The weight of term placenta is 500–600 g, the estimated surface area of syncytiotrophoblast available for maternofetal transport is 11–13 m^2, and the inner fetal capillary surface area is about 12 m^2 [22].

Human placenta is formed by chorionic plate and its repeatedly branched projections, chorionic villi, that are immersed in maternal blood circulating in the intervillous space. The growth of villi goes hand in hand with fetal growth. They develop into various types during pregnancy, and their classification depends on size and structural features. Basically, each villus is covered by the layer of trophoblast consisting of continuous cytoplasmic mass with numerous nuclei, that is, syncytiotrophoblast, and cuboidal cells of cytotrophoblast appearing as a continuous layer underneath syncytiotrophoblast in early gestation, but sparsely spread in term placenta. As nuclei in syncytiotrophoblast do not divide, cytotrophoblast cells play the role of stem cells of trophoblast. They undergo mitotic division, and their fusion with syncytiotrophoblast enlarges its mass. As shown by detection of cell cycle markers (e.g., Ki67, PCNA), they display proliferative potential over the duration of pregnancy [23–25].

The other source of placental tissues is extraembryonic mesoderm. For the placental development, this cell population gives rise to the mesenchyme adjoining villous trophoblast. Mesenchymal derivatives, that is, cells of connective tissue, endothelial cells, pericytes, and smooth muscle cells then form villous stroma and vascular bed. It is obvious that they follow the growth of trophoblast, and their proliferation and subsequent differentiation contribute to the formation of new functionally efficient villi, and thus to the enlargement of the organ and its functional capacity. It is significant particularly in the third trimester when the rapid development of terminal villi accompanies the rapid enlargement of fetal size [22]. Previous studies have demonstrated the proliferative potential of cells in villous vascular bed and stroma in early pregnancy as well as at term [24, 25]. The proliferative potential of cells in terminal villi of normal human term placenta is also demonstrated in **Figure 1**.

In order to provide nutrition for the increasing metabolic requirements of growing fetus, the placenta displays continuous cell proliferation during its entire existence. Despite the fact that fetal demands attain their maximal level at term, it is expected that cells in placenta at term decrease their proliferative capacity due to telomere shortening and undergo senescence similar to somatic cells in other organs. It is also supposed that those changes in telomere length might take part in the initiation of parturition [26]. Telomere shortening during pregnancy was found also in the placenta of mice [27]. As shown in another study in mouse, the proportion of representative short telomeres, that is, 3- and 5-kb telomere fragments,

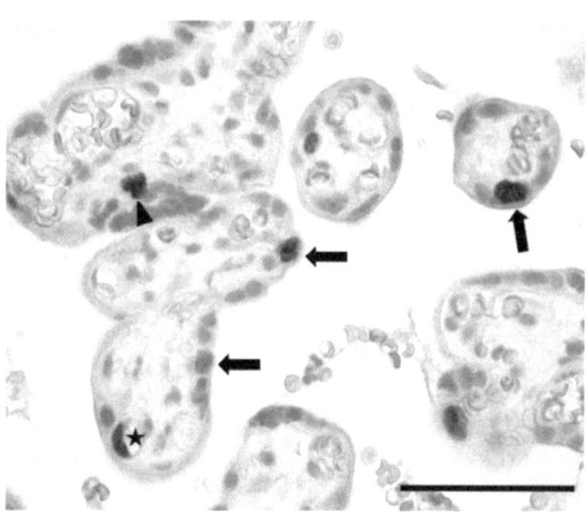

Figure 1.
Proliferative activity of cytotrophoblast (arrows), stromal cell (arrowhead), and capillary endothelium (asterisk) in terminal villi of normal term placenta demonstrated by Ki-67 immunohistochemistry. Bar = 50 μm.

significantly increases toward the end of pregnancy. The authors conclude that the quantity of representative short telomeres increases prior to parturition and takes part in the mediation of cellular aging in the placenta, finally leading to parturition [26]. Those findings are consistent with other studies regarding the impact of short telomere rather than average telomere length [28].

Due to its location in maternal uterus, the placenta is influenced by any abnormal metabolic conditions in both mother and fetus. The negative influence demonstrates itself in pathological structural features as well as in impaired function. In such organ displaying mitotic activity, altered telomere homeostasis may take part in those adverse changes. Here, we summarize data on telomere length and telomerase activity in placentas from normal pregnancies and pregnancies complicated by metabolic disturbances threatening pregnancy outcome, as are maternal diabetes mellitus, preeclampsia, and intrauterine growth restriction.

3. Analysis of telomere homeostasis—methodological approaches

In view of the facts mentioned above, telomere length emerged as a promising marker generally in medicine, but the utility of such measurements highly depends on valid methodologies [7]. The average telomere length is highly variable among different cell types and among different individuals. There are also gender differences, which could be detected at birth. Additionally, the average telomere length declines with age [16].

Telomeres may be repaired either by the enzyme telomerase using its RNA template or by recombination. The methodologies examining telomere homeostasis in different tissues are therefore based not only on the determination of the length of telomeric DNA sequences, but they examine also the status and expression of the reverse transcriptase *(hTERT)* and the RNA template *(hTERC)* genes and recombination events on selected telomeres. In the following paragraphs and in **Table 1**, we summarize the main features and limitations of methodologies employed in placenta research, alternatively the methods applicable on human samples.

3.1 Methods examining telomeric DNA sequences

Terminal restriction fragment (TRF) analysis was originally developed to determine the lengths of telomeres in 1988 by Moyzis et al. [54]. Genomic DNA is isolated and then digested with a mixture of restriction enzymes that are selected to avoid the cutting of telomeric and subtelomeric sequences. The DNA fragments obtained after such a digestion are separated in agarose gel and hybridized with the probe containing telomeric sequences [55]. The need for high amounts of highly integral DNA samples represents one of the main limitations of this methodology [7].

Polymerase chain reaction (PCR)-based methodologies were elaborated to overcome limits of the previous method. Cawthon [56] reported the technique based on quantitative PCR (qPCR). The sequences of primer pairs proposed by Cawthon avoid the primer dimer formation which frequently occurs when repetitive sequences are amplified. Cawthon's qPCR technique and its subsequent modifications [56–58] belong to the most frequently used methods. Telomere length is quantified relatively—the ratio T/S between the quantity of repetitive telomeric sequences (T) and the quantity of sequences representing a single copy gene (S) is calculated. These ratios correlated very well with the absolute telomere lengths measured by TRF method in Cawthon's original study [56]. An attempt to standardize this methodology over large spectrum of laboratories was made [59].

Method	References
Telomere length analysis based on isolated DNA	
TRF	[29–31]
STELA	[32–34]
Q-PCR, T/S ratio	[29, 35–38]
Telomere length analysis based on examination of interphase nuclei	
Interphase Q-FISH	[36, 39–45]
Determination of gene copy numbers in situ	
hTERT-FISH	[36]
hTERC FISH	[36, 39, 41, 46]
hTERT expression	
Immunohistochemistry	[43, 45, 47]
RT-PCR for hTERT	[29, 41, 48–51]
hTERC expression	
RT-PCR for hTERC	[29]
Telomerase activity determination	
TRAP	[33, 38, 48, 52, 53]
TRAP *in situ*	[52, 53]
Genome-wide technologies	
Illumina methylation array	[37]

Table 1.
Experimental methods used in placenta research.

O'Callaghan and Fenech [60] described the modification of qPCR-based technique allowing the determination of an absolute telomere length due to the use of artificially synthesized DNA standard containing the defined number of telomeric repeats.

PCR-based techniques are quite popular among researchers because they require lower DNA quantity and usual laboratory equipment. Due to their relatively low costs, they may be employed in high-throughput epidemiologic studies. One shortcoming of PCR-based methodologies is that the resulting values inform only about the average telomere length in the specimens.

Single telomere length analysis (STELA) targets telomeric DNA from a single chromosome using primers specific to subtelomeric sequences [61]. All individual chromosomal subtelomeric regions do not contain suitable specific sequences; therefore, the method is able to examine only a limited set of chromosomes [7]. Using this methodological approach, Garcia-Martin et al. [32] found considerable intra-sample variability in examined placentas.

Quantitative fluorescence in situ hybridization (Q-FISH)-based techniques are focused more on the work with cells than with isolated DNA. A fluorescent probe, mostly peptide nucleic acid (PNA) probe, complementary to telomeric repetitive sequences is hybridized to denatured DNA of metaphase chromosomes or interphase nuclei. The cells may be fresh, frozen, or formaldehyde-fixed and paraffin-embedded. The application of *metaphase Q-FISH*, which was developed in 1996 by Lansdorp and colleagues [62], results in the estimation of the length of all 92 telomeres in human cells. It allows the detection of telomere free ends. The most serious limitation of this approach is the requirement of metaphases from mitotically active cells. The method is labor intensive [7, 62].

Interphase Q-FISH, first described in 1998 by de Pauw and colleagues [63], is applicable on nondividing cells. It compares the fluorescent signals obtained after hybridization with a telomere-specific probe and with a probe targeting a single copy gene. Its results inform, similar to the results of qPCR methodologies, about the average length of the telomeres in examined cell, the method is not able to recognize each individual telomere as metaphase Q-FISH.

A method for telomere capture evaluation based on interphase Q-FISH has been established by Amiel et al. [64]. When telomeres shorten to the critical length, repair pathways are activated. In the process of telomere capture, a critically short telomere obtains a new telomeric sequence from another chromosomal end. In the original method, the number of fluorescent signals for a single copy gene, SNRP, which is localized on chromosome 13, was compared with the number of signals for 15qter region of this chromosome [64] to follow not only random aneuploidy but also telomere capture or translocation of telomere. The methodology was also used in placenta research [39, 48].

The *pq-ratio assay* described in 2001 by Perrem et al. [40] belongs to methods that directly examine telomere recombination. This assay measures the variation in telomere lengths at the p and q arms of a chromosome. The telomere ratio for most chromosomes is expected to be q/p~1, because the telomeres at both ends of a given chromosome shorten at a similar rate. If recombination is used to maintain the ends, then it could alter the length of at least one telomere by a random amount of telomeric repeats. This results in variable values for the pq-ratio. The pq-ratio assay is very sensitive, and the data may become biased as the telomeres shorten. Small changes on a short telomere may be overrepresented and telomeres with no signal will not be represented at all [40, 65, 66].

Chromosome orientation-FISH (CO-FISH) was first described in 1993 [67, 68] as a method for strand-specific FISH. The method is dependent on cultivation of analyzed cells because it requires incorporation of bromodeoxyuridine (BrdU) into newly synthesized strands. This step allows subsequent enzymatic removal of BrdU containing strands after their damage caused by UV light in the presence of the dye Hoechst 33258. The remaining strands then serve as single-stranded targets for FISH. This approach allows differentiation between the telomeres produced via leading- or via lagging-strand DNA synthesis, and it enables the study of sister chromatid exchanges (SCE) and inversions in telomeric regions [69, 70].

The methodology of *directional genomic hybridization (dGH)* represents the cytogenomic extension of strand-specific hybridization. *Telo-dGH* recognizes terminal exchange events—terminal inversions and generally different forms of genetic recombination occurring near the telomeres, namely sister chromatid exchange (SCE) [71].

Flow-FISH represents the modification of interphase Q-FISH, which was introduced in 1998 by Hultdin et al. [72]. The cells in suspension are hybridized with fluorescent probes and then examined using flow cytometry. This approach is technically very demanding because the unfixed cells are often fragile and clustering. The technique is very sensitive to preservation of cells. FISH probes may also have affinity to cytoplasmic structures [7]. Higher numbers of cells are needed (typically $> 1 \times 10^5$), the assay determines mean telomere lengths and it does not account for aneuploidy or SFE (signal free ends) but the methodology was successfully used in numerous studies [73–75].

TELI-FISH is a combined FISH/immunofluorescence method which was developed in 2002 to assess human telomere lengths from standard formalin-fixed paraffin-embedded tissues. Combination with immunostaining allows the simultaneous identification of specific cell types. The assay requires very few cells (10–15).

Validation showed excellent agreement with the commonly used TRF method based on Southern blotting [76].

Microdissection followed by qPCR allows selection of cells for analysis using classical histological technique and it is suitable for the studies based on archival material [35]. Microdissected cells may be examined by qPCR methodology, and therefore, its main disadvantage is that it provides information only about the relative average length of telomeres in analyzed samples. The methodology does not result in the determination of the lengths of the longest or the shortest telomeres contained in the sample. Comparison of T/S ratios related to a reference sample allows evaluation of differences between the groups of samples (for instance between healthy controls and patients).

Whole genome sequencing (WGS) captures sequence information from the entire genome, including the telomeres, and is increasingly being applied in research and in the clinic. In 2014, Ding et al. [77] demonstrated a novel method, TelSeq, which allows measurement of average telomere length by using whole genome or exome sequencing data. It was the first study that evaluated in detail the relationship between the frequency of telomere repeats and telomere length. With the potential to be a relatively high-throughput method, this may overtake qPCR as the method of choice in future studies. Their study was the first computational method that had been validated against an established experimental method (Southern blot measurements of the mean length of terminal restriction fragments).

3.2 Methods examining genes coding for telomerase and its RNA components, their expression, and telomerase activity

The telomere length is closely associated with telomerase (human telomerase reverse transcriptase—hTERT) activity (TA) and the availability of its RNA component (hTERC or TERRA—telomeric repeat-containing RNA) in tissues.

The copy number of both genes hTERT and hTERC in nuclei in archive tissues or cultivated cells is examined by *FISH methodology*. This approach was applied also in placenta research [36, 41, 46, 48]. *RNA-FISH* is based on the use of fluorescently labeled probes and allows the cellular localization of TERC [78].

The presence of telomerase itself may be detected in tissues by means of *immunohistochemistry*—a methodology which is generally well established in laboratories of pathologists [76]. This methodology was also widely used in placenta research—see reference in **Table 1**.

In **Table 1**, the studies that examined the expression of *hTERT* using *reverse transcription and subsequent qPCR* to quantify the amount of *hTERT* transcripts are also summarized.

The telomerase activity may be measured by a wide panel of methodologies based on addition of telomerase substrate and detection of amplified telomerase products by *telomere repeat amplification protocols—TRAPs*—developed by Kim et al. [79]. Recently existing numerous methodological modifications of this approach were reviewed by Mensa et al. [78].

hTERC (TERRA) is a long noncoding RNA, which can be transcribed from nearly all telomeres in mammalian cells because its transcription starts from their subtelomeric regions. Therefore, the quantity of such transcripts can be measured by *RT-qPCR* starting from RNA isolated from analyzed cells or tissues followed by reverse transcription and PCR with chromosome-specific primers. Molecular mechanisms associated with the role of TERC in telomere reparation are intensively studied [80, 81].

Northern blotting or *RNA dot blotting* needs mostly radiolabeled probes and cannot detect the minor changes in quantity of TERC [78].

Detection of G4 quadruplexes using antibodies is based on the fact that during hTERC transcription, the RNA:DNA hybrids at the chromosome ends are formed, and they can fold into G-quadruplexes [78]. The G-quadruplex structure formed by telomere DNA plays also an important role in the regulation of the telomerase reaction [82].

3.3 Methods examining epigenetic changes and chromatin structure

Not only the presence of DNA sequences of genes *hTERT* and h*TERC* but also their functional state determined epigenetically plays a crucial role in the regulation of their expression.

Wilson et al. [37] used *array technology* (Illumina Infinium Human Methylation 450 BeadChip) to study methylation alterations in genes *hTERT, DNMT1,* and *DNMT3A* in human placentas. *MALDI-TOF technology* was employed to determine the level of DNA methylation of *hTERC* in placenta using Sequenom EpiTyper platform [29]. Lower levels of gene methylation were found in normal placentas compared with other somatic cells.

Quantitative Telomeric Chromatin Isolation Protocol (QTIP) was introduced in 2013 by Grolimund et al. [83]. It allows the comprehensive determination of telomere protein composition and the quantitative comparison of telomere protein compositions between cells with different telomeric states. Chromatin is cross-linked, immunopurified, and analyzed by mass spectrometry. The methodology may be also adapted for examination of other chromatin regions within the genome [83].

4. Placenta in normal pregnancy

The assessment of telomere length in normal pregnancy gives important data regarding dynamics of placental cellular proliferation. The study on third trimester placentas using the qPCR has shown decreased telomere length between gestational weeks 28 and 42 (13.98–10.56 kbp) [38]. The application of qPCR and Southern blot-based terminal restriction fragment (TRF) assay confirmed considerably longer telomeres in first trimester villi than in term placentas, telomeres of which were found to be longer than those in cord blood mononuclear cells [29, 84].

Some authors took into consideration that the position of villous tissue in the placenta may influence the telomere length, and therefore collected and processed samples of the whole placenta from more locations [29, 37, 84]; nevertheless, no site-specific differences of telomere length were determined except [84] showing that the telomerase activity was detected in term placenta restricted to biopsy sites near umbilical cord only. Moreover, the Southern blot-based TRF assay discovered longer telomere length in placental samples than in cord blood cells [84]. The study by qPCR proved that telomeres in placentas of female fetuses are longer than in placentas of male fetuses at the same gestational week. This finding suggests an influence of hormonal milieu during intrauterine development [37]. On the other hand, the other study performed by single telomere length analysis (STELA) revealed neither influence of fetal sex nor influence of the mode of delivery [32].

The above-mentioned findings in chorionic villi and normal placenta suggest that telomere length is maintained by active telomerase during pregnancy. Nevertheless, the papers dealing with this topic present equivocal results. Using TRAP assay, Wright et al. [33] found no detectable telomerase activity in placenta. To the contrary, telomerase activity studied by TRAP assay and in situ TRAP assay was found in both chorionic villi at 5–14 weeks of gestation and normal placentas at 23–42 weeks of gestation [49]. Expression of telomerase protein was detected by immunohistochemistry in cytotrophoblast of chorionic villi [47, 85]. Decreasing relative telomerase activity was shown in comparison of chorionic villi in first, second and third trimester. In normal villi from 6 to 40 gestational weeks, the presence of telomere RNA component (TERC) was demonstrated, whereas telomerase reverse transcriptase (TERT) was not found in samples from second and third trimesters [86]. Immunohistochemical reaction revealed telomerase expression in trophoblast and stroma of villi in 10th week, but not in term placenta [87]. Using RT-PCR method, the hTERT-RNA expression was found in normal chorionic villi from 6th to 10th gestational week as well as in normal placentas from 12th to 41st gestational week, whereas hTERT protein expression was found in chorionic villi, but only in the fourth part of placental samples [88].

5. Placenta in pregnancy pathologies

It is evident that normal placental growth and development carried out by cell proliferation is conditioned by appropriate telomere length ensured by homeostasis of telomerase system. For optimal course of those processes, normal metabolic milieu is necessary in mother, placenta, and fetus. The most critical condition of intrauterine development is adequate oxygen supply. The early pregnancy phase, that is, first 10 weeks, runs under low oxygen levels. This relative hypoxia before the constitution of the fetoplacental and uteroplacental blood circulation induces various factors, one being the hypoxia-inducible factor 1, that upregulates hTERT expression (and telomerase activity), and its decrease with gestational age is in correlation with decrease of telomerase activity logically followed by telomere shortening [88]. As the placenta consumes about 40% of the oxygen supplied to fetoplacental unit [89], the hypoxic conditions have negative impact on all processes running there. And hypoxia also represents a key factor in genesis of pregnancy pathologies discussed in the following parts of this chapter.

5.1 Maternal diabetes mellitus

There are two main forms of maternal diabetes mellitus, the insulin-dependent form with onset before conception, and gestational diabetes diagnosed usually in second half of pregnancy that disappears after birth. In the insulin-dependent type diabetes, an autoimmune process destroys β-cells of the islets of Langerhans completely and the patient is then treated by insulin supplementation.

As shown in experiments with mice, the type 2 diabetes, and similarly gestational diabetes, may involve telomere shortening during pathogenesis. Shorter telomeres are associated with impaired β-cell regeneration, impaired glucose-stimulated insulin secretion by disorders of insulin release leading to impaired glucose tolerance as well as to increased β-cell senescence [14, 90].

Both forms of maternal diabetes are characterized by maternal hyperglycemia and thus higher amount of glucose transported to fetus. Metabolic complications derived from hyperglycemia threaten the mother; manifest themselves in placental structure and function; have negative influence on fetal well-being, perinatal morbidity and mortality; and long-lasting effect on the postnatal life of the individual. In pregnancies complicated by maternal diabetes mellitus, the alterations of placental structure and function as well as the pregnancy outcome depend on the quality of metabolic control. Placentas from poorly controlled diabetes are larger and heavier and microscopic picture shows disturbances of villous maturation [22]. The oxidative stress produced by imbalance of glucose and oxygen supply in fetoplacental unit is compensated by enhanced placental angiogenesis demonstrated by higher villous capillary branching [91]. It is possible to suppose that both, the higher placental weight and enhanced angiogenesis, are a consequence of escalated cell proliferation leading to exhaustion of telomeres available for mitotic division.

Regarding the telomere length in placenta from pregnancies complicated by maternal diabetes, the available data are not quite consistent. In the qPCR-based study examining large cohort of placentas, maternal diabetes was found associated with longer telomere length [38]. In another study comparing telomere length in cytotrophoblast of placentas from poorly controlled maternal diabetes and normal pregnancies by FISH method, the result indicated shorter telomeres in the diabetic group [42]. In the same groups of patients, the immunohistochemically identified telomerase expression, the expression of mRNA for hTERT, and the expression of *TERC* gene copy number were lower in diabetic placentas [41]. On the other hand, no difference of mean telomere length was found in peripheral villi of normal placentas and placentas in well-controlled maternal diabetes examined by laser capture microdissection and qPCR [35], although the study on the same placental material discovered lower proliferative potential of cytotrophoblast and vascular endothelium of terminal villi in maternal diabetes [25]. Single telomere length analysis (STELA) was used for measurement of telomere length in normal placentas and placentas from gestational diabetes treated either by lifestyle intervention or by metformin or insulin therapy. The result showed that the therapy by metformin or insulin protected from telomere shortening in placentas of male fetuses [34].

5.2 Preeclampsia

Preeclampsia is a disease of pregnancy characterized by new-onset maternal hypertension and proteinuria. It may begin in 28–34 gestational weeks (early-onset preeclampsia) or after 34 weeks (late-onset preeclampsia). It is commonly accepted that the preeclampsia originates in deficient placentation, that is, decreased invasion of the maternal tissues by extravillous trophoblast. Under normal conditions, the endothelium and smooth muscle cells of uterine spiral arterioles are replaced with trophoblast and their diameter becomes wide allowing delivery of blood at low pressure to the intervillous space. The decreased trophoblastic invasion produces narrow uteroplacental arteries and the resulting malperfusion of the intervillous space causes oxidative stress of the fetoplacental unit manifested among others by reduced development of the villous tree and placental growth retardation. Increased placental proliferative activity found using detection of PCNA and Ki67 in preeclampsia may be a sign of increased cell turnover [92]. The associated systemic vascular inflammation in maternal organism may cause injury of multiple organs.

In placentas in preeclampsia, the analysis of villous cytotrophoblast based on the quantitative FISH method showed shorter telomeres, more end-to-end telomere aggregates, and abnormal *TERC* gene copy number as well as decreased hTERT expression detected by immunohistochemistry [39, 43, 44]. Common expression of hTERT protein and HIF-1α in term preeclamptic placenta gives an evidence of response to hypoxia by telomerase upregulation [88]. Nevertheless, the measurement of average telomere length by qPCR did not show significant differences between control, early-onset preeclamptic, and later-onset preeclamptic placentas [37]. Another study has also shown no differences of telomere length between normal placentas and placentas in preeclampsia [30]. Enhanced levels of placental hTERT-mRNA in preeclampsia [50] if not associated with longer telomeres as a result of enhanced telomerase activity suggests that there is a conceivable disturbance in translation or post-translation processes of the enzyme protein.

6. Intrauterine growth restriction

In intrauterine growth restriction (IUGR), the growth and development of fetus is delayed by 3–4 weeks regarding the gestational age and the birth weight is low, under 10th percentile for gestational age. Fetal growth restriction is associated with restricted placental size caused by arrest mechanisms reducing cell proliferation [45]. There are two types of IUGR, symmetric and asymmetric. The fetus displaying symmetric IUGR has normal body proportion, the fat and muscle tissue are reduced. It is usually associated with genetic factors causing, for example, already mentioned bone marrow syndromes [9–12] or infections. The asymmetric IUGR is characterized by normal size of head and reduced chest and abdominal circumference due to reduced fat and muscle tissue. It is often associated with placental insufficiency arising, for example, in preeclampsia and may be related to oxidative stress.

Studies performed by quantitative FISH method and RT-PCR on placental cytotrophoblast in IUGR gave an evidence of shorter telomeres, lower telomerase activity, decreased hTERT mRNA, and decreased *TERC* gene copy number [45, 46, 48]. The relative telomere length and hTERT expression were found lower in cytotrophoblast of placentas in IUGR as well as in IUGR combined with preeclampsia [43].

Lower proliferative potential found in placenta in IUGR [52, 70, 76, 93] seems to be consistent with decreased telomerase activity in cytotrophoblast of IUGR placenta [52, 76]. In placenta associated with asymmetric IUGR, only weak, if any, telomerase activity, hTERT expression, and copy numbers of telomerase reverse transcriptase were found by qPCR and in situ TRAP assay [49, 51, 53]. Shorter telomeres associated with higher expression of cell senescence markers were found in placenta samples in IUGR [31] and shorter telomeres detected by quantitative FISH technique and reduced average telomere length detected by qPCR were shown in [36]. To the contrary, no difference was revealed in average telomere length assessed by qPCR between normal placenta and placenta in IUGR [37].

7. Conclusions

Appropriate fetal growth and development is conditioned by appropriate placental growth and development. It is accomplished by balanced cell proliferation, differentiation, and apoptosis. Cell proliferation is influenced by the length of telomeric sequences of chromosomes and their elongation due to telomerase activity. This review article summarizes available data on telomere length and telomerase activity of placenta in pregnancy-complicating situations, that is, maternal diabetes mellitus, preeclampsia, and intrauterine growth restriction (**Table 2**) as well as methods used for this research (**Table 1**).

In normal placentas, longer telomeres and higher telomerase activity were found in early pregnancy, they gradually decreased till the term. Although the available studies on placentas from pregnancies complicated by maternal diabetes are not numerous, their results corroborated by experimental studies suggest that diabetic metabolic conditions contribute to telomere shortening and that the appropriate metabolic control achieved by adequate treatment may function as a prevention of this adverse process. The results of research on telomere length and telomerase activity in preeclampsia are still equivocal and rather suggest debatable comparability of methods applied in those studies. The telomere length in placenta associated with IUGR was found lower and accompanied with decreased expression and activity of components of telomerase apparatus.

Diagnosis	Telomere length	Telomerase activity
Normal placenta	Telomere length is reduced in the course of pregnancy; placental telomere length is not site-specific [32, 33, 37, 38, 84]	hTERT expression and telomerase activity were found [47, 49, 84, 85]; telomerase activity decreases in the course of pregnancy [86–88]
Placenta associated with maternal diabetes mellitus	Maternal diabetes is associated with longer placental telomeres [38]; there is no difference in placental telomere length in well-controlled diabetes [34, 35]; telomeres are shorter in placentas from poorly controlled diabetes [42]	Lower expression of hTERT, hTERT-mRNA, lower *TERC* copy number in placentas from poorly controlled diabetes [41]
Placenta associated with preeclampsia	No differences of telomere length between normal placentas and placentas in preeclampsia [37, 84]; shorter telomeres, more end-to-end telomere aggregates, higher telomere aggregates count [43, 44]	Lower expression of hTERT [44], abnormal *TERC* copy gene number [39]; higher expression of hTERT mRNA [50]
Placenta associated with IUGR	Shorter telomeres [31, 36, 45]; no difference in telomere length [37]	Lower expression of hTERT, hTERT-mRNA, lower *TERC* copy number [46, 48, 51–53]

Table 2.
Summarized data on telomere length and telomerase activity in normal placenta and placenta in case of maternal diabetes mellitus, preeclampsia, and IUGR.

Maternal diabetes mellitus, preeclampsia as well as IUGR do not only complicate pregnancy but are also taken as causes of adverse outcomes for individuals later in postnatal period. Further investigation of the effect of factors influencing telomere length and telomerase activity may contribute to better understanding of those links. The placental DNA is identical to the DNA of fetal cells and the period of its existence equal to the prenatal period of the fetus spent in the same maternal environment. As telomeres are susceptible to external conditions of maternal milieu (i.e., oxidative stress, reactive oxygen species, exogenous genotoxic insults), those epigenetic influences may accelerate their shortening [94]. Prospective studies in experimental models and of course in long-time prospective studies in human should elucidate if there is a relation of final telomere length in placentas at term and newborn and if the measurement of placental telomere length could have a predictive potential for individual.

Recently, the immunomodulatory role of telomeric sequences was recognized [78, 95]. Telomeric sequences originated from trophoblasts may circulate in the pool of cell-free DNA in maternal plasma and contribute to timing of parturition [26] by stimulation of maternal immune response against placenta. The role of telomeric sequences contained in cell-free DNA in plasma of healthy persons in the regulation of immune system performance was also described [95]. Additionally, the telomeric sequences were found also in cytoplasm where they regulate inflammatory response via their interaction with TLR9 receptor [78].

In view of all these facts, the study of telomeres and their homeostasis in placenta seems to be crucial for the understanding of pathogenesis in the broad spectrum of pregnancy complications.

Acknowledgements

This work was funded by the grant No. Progres Q25/LF1 of the Ministry of Education, Youth and Sport of the Czech Republic, and by the grant RVO-VFN 64165 of the Ministry of Health of the Czech Republic.

DOI: http://dx.doi.org/10.5772/intechopen.86327

Conflict of interest

There are no conflicts of interest.

Abbreviations

FISH	fluorescence in situ hybridization
IUGR	intrauterine growth restriction
HIF	hypoxia induced factor
hTERT	human telomere reverse transcriptase = catalytic component of telomerase activity
IUGR	intrauterine growth retardation
qPCR	quantitative polymerase chain reaction
RT-PCR	reverse transcriptase polymerase chain reaction
SFE	signal free ends
TERC	telomerase RNA component gene
TERT	telomerase catalytic component gene
TRAP	telomeric repeat amplification protocol
TRF	terminal restriction fragment

Author details

Marie Jirkovská*, Marie Korabečná and Soňa Laššáková
First Faculty of Medicine, Charles University, and General University Hospital in Prague, Prague, Czech Republic

*Address all correspondence to: mjirk@lf1.cuni.cz

References

[1] Keefe DL. Telomeres and genomic instability during early development. European Journal of Medical Genetics. 2019:S1769-7212(19)30058-8. DOI: 10.1016/j.ejmg.2019.03.002. [Epub ahead of print]

[2] Ozturk S, Sozen B, Demir N. Telomere length and telomerase activity during oocyte maturation and early embryo development in mammalian species. Molecular Human Reproduction. 2014;**20**:15-30. DOI: 10.1093/molehr/gat055

[3] Wright DL, Jones EL, Mazer JF, Oehninger S, Gibbons WE, Lanzendorf SE. Characterization of telomerase activity in the human oocyte and preimplantation embryo. Molecular Human Reproduction. 2017;**10**:947-955

[4] Liu L, Bailey SM, Okuka M, Muñoz P, Li C, You L, et al. Telomere lengthening early in development. Nature Cell Biology. 2007;**9**:1436-1441

[5] Lai SR, Phipps SM, Liu L, Andrews LG, Tollefsbol TO. Epigenetic control of telomerase and modes of telomere maintenance in aging and abnormal systems. Frontiers in Bioscience. 2005;**10**:1779-1796

[6] Hayflick L, Moorhead PS. The serial cultivation of human diploid cell strains. Experimental Cell Research. 1961;**25**:585-621

[7] Montpetit AJ, Alhareeri AA, Montpetit M, Starkweather AR, Elmore LW, et al. Telomere length: A review of methods for measurement. Nursing Research. 2014;**63**:289-299. DOI: 10.1097/NNR.0000000000000037

[8] López-Otín C, Blasco MA, Partridge L, Serrano M, Kroemer G. The hallmarks of aging. Cell. 2013;**153**: 1194-1217. DOI: 10.1016/j.cell.2013.05.039

[9] Theimer CA, Finger LD, Trantirek L, Feigon J. Mutations linked to dyskeratosis congenita cause changes in the structural equilibrium in telomerase RNA. Proceedings of the National Academy of Sciences. 2003;**100**: 449-454. DOI: 10.1073/pnas.242720799

[10] Yamaguchi H, Calado RT, Ly H, Kajigaya S, Baerlocher GM, Chanock SJ, et al. Mutations in TERT, the gene for telomerase reverse transcriptase, in aplastic anemia. The New England Journal of Medicine. 2005;**352**:1413-1424

[11] Brümmendorf TH, Balabanov S. Telomere length dynamics in normal hematopoiesis and in disease states characterized by increased stem cells turnover. Leukemia. 2006;**20**:1706-1716

[12] Boddu PC, Kadia TM. Molecular pathogenesis of acquired aplastic anemia. European Journal of Haematology. 2019;**102**:103-110. DOI: 10.1111/ejh.13182

[13] McDonough JE, Martens DS, Tanabe N, Ahangari F, Verleden SE, Maes K, et al. A role for telomere length and chromosomal damage in idiopathic pulmonary fibrosis. Respiratory Research. 2018;**19**:132. DOI: 10.1186/s12931-018-0838-4

[14] Guo N, Parri EM, Li L-S, Kembou F, Lauder N, Hussain MA, et al. Short telomeres compromise β-cell signalling and survival. PLoS One. 2011;**6**:e17858. DOI: 10.1371/journal.pone.0017858

[15] Giri N, Reed HD, Stratton P, Savage SA, Alter BP. Pregnancy outcomes in mothers of offspring with inherited bone marrow failure syndromes. Pediatric Blood & Cancer. 2018;**65**:e26757. DOI: 10.1002/pbc.26757

[16] Aubert G, Baerlocher GM, Vulto I, Poon SS, Lansdorp PM. Collapse of telomere homeostasis in hematopoietic calls caused by heterozygous mutations in telomerase genes. PLoS Genetics. 2012;**8**:e1002696. DOI: 10.1371/journal.pgen.1002696

[17] Shay JW, Wright WE. Telomeres and telomerase: Three decades of progress. Nature Reviews. Genetics. 2019;**20**:299-309. DOI: 10.1038/s41576-019-0099-1

[18] Marrone A, Stevens D, Vulliamy T, Dokal I, Mason PJ. Heterozygous telomerase RNA mutations found in dyskeratosis congenita and aplastic anemia reduce telomerase activity via haploinsufficiency. Blood. 2004;**104**:3936-3942

[19] Armanios M, Chen JL, Chang YP, Brodsky RA, Hawkins A, Griffin CA, et al. Haploinsufficiency of telomerase reverse transcriptase leads to anticipation in autosomal dominant dyskeratosis congenita. Proceedings of the National Academy of Sciences of the United States of America. 2005;**102**:15960-15964

[20] Goldman F, Bouarich R, Kulkarni S, Freeman S, Du HY, Harrington L, et al. The effect of TERC haploinsufficiency on the inheritance of telomere length. Proceedings of the National Academy of Sciences of the United States of America. 2005;**102**:17119-17124

[21] Mayhew TM, Wadrop E, Simpson RA. Proliferative versus hypertrophic growth in tissue subcompartments of human placental villi during gestation. Journal of Anatomy. 1994;**184**:535-543

[22] Benirschke K, Kaufmann P. Pathology of the Human Placenta. 3rd ed. New York: Springer Verlag; 1995

[23] Danihel L, Gomolčák P, Korbel M, Pružinec J, Vojtaššák J, Janík P, et al. Expression of proliferation and apoptotic markers in human placenta during pregnancy. Acta Histochemica. 2002;**104**:335-338. DOI: 10.1078/0065-1281-00683

[24] Korgun ET, Celik-Ozenci C, Acar N, Cayli S, Desoye G, Demir R. Location of cell cycle regulators cyclin B1, cyclin A, PCNA, Ki67 and cell cycle inhibitors p21, p27 and p57 in human first trimester placenta and decidua. Histochemistry and Cell Biology. 2006;**125**:615-624. DOI: 10.1007/s00418-006-0160-y

[25] Jirkovská M, Kučera T, Dvořáková V, Jadrníček M, Moravcová M, Žižka Z, et al. Impact of maternal diabetes type 1 on proliferative potential, differentiation and apoptotic activity in villous capillaries of term placenta. Placenta. 2016;**40**:1-7. DOI: 10.1016/j.placenta.2016.02.003

[26] Phillippe M, Sawyer MR, Edelson PK: The telomere gestation clock: Increasing short telomeres at term in the mouse. American Journal of Obstetrics and Gynecology 2019; pii: S0002-9378(19)30248-0.DOI:10.1016/j.ajog.2019.01.218. [Epub ahead of print]

[27] Bonney EA, Krebs K, Saade G, Kechichian T, Trivedi J, Huaizhi Y, et al. Differential senescence in feto-maternal tissues during mouse pregnancy. Placenta. 2016;**43**:26-34. DOI: 10.1016/j.placenta.2016.04.018

[28] Hemann MT, Strong MA, Hao LY, Greider CW. The shortest telomere, not average telomere length, is critical for cell viability and chromosome stability. Cell. 2001;**107**:67-77

[29] Novakovic B, Napier CE, Vryer R, Dimitriadis E, Manuelpillai U, Sharkey A, et al. DNA methylation mediated up-regulation of *TERRA* non-coding RNA is coincident with elongated telomeres in the human placenta.

Molecular Human Reproduction. 2016;**22**:791-799. DOI: 10.1093/molehr/gaw053

[30] Broady AJ, Loichinger MH, Ahn HJ, Davy PMC, Allsop RC, Bryant-Greenwood GD. Protective proteins and telomere length in placentas from patients with pre-eclampsia in the last trimester of gestation. Placenta. 2017;**50**:44-52. DOI: 10.1016/j.placenta.2016.12.018

[31] Davy P, Nagata M, Bullard P, Fogelson NS, Allsop R. Fetal growth restriction is associated with accelerated telomere shortening and increased expression of cell senescence markers in the placenta. Placenta. 2009;**30**:539-542. DOI: 10.1016/j.placenta.2009.03.005

[32] Garcia-Martin I, Janssen AB, Jones RE, Grimstead J, Penketh RJA, Baird DM, et al. Telomere length heterogeneity in placenta revealed with high-resolution telomere length analysis. Placenta. 2017;**59**:61-68. DOI: 10.1016/j.placenta.2017.09.007

[33] Wright WE, Piatyszek MA, Rainey WE, Byrd W, Shay JW. Telomerase activity in human germline and embryonic tissues and cells. Developmental Genetics. 1996;**18**:173-179. DOI: 10.1002/(sici)1520-6408(1996)18:2<173::aid-dvg10>3.3.co;2-5

[34] Garcia-Martin I, Penketh RJA, Janssen AB, Jones RE, Grimstead J, Baird DM, et al. Metformin and insulin treatment prevent placental telomere attrition in boys exposed to maternal diabetes. PLoS One. 2018;**13**:e0208533. DOI: 10.1371/journal.pone.0208533

[35] Zinková A, Marová D, Koperdáková J, Mirchi TP, Korabečná M, Jirkovská M. Relative amount of telomeric sequences in terminal villi does not differ between normal term placentas and placentas from patients with well-controlled type 1 diabetes mellitus. Placenta. 2017;**55**:1-4. DOI: 10.1016/j.placenta.2017.04.016

[36] Toutain J, Prochazkova-Carlotti M, Cappellen D, Jarne A, Chevret E, Ferrer J, et al. Reduced placental telomere length during pregnancies complicated by intrauterine growth restriction. PLoS One;**8**(1):e54013. DOI: 10.1371/journal.pone.0054013

[37] Wilson SL, Liu Y, Robinson WP. Placental telomere length decline with gestational age differs by sex and TERT, DNMT1, and DNMT3A DNA methylation. Placenta. 2016;**48**:26-33. DOI: 10.1016/j.placenta.2016.10.001

[38] Gielen M, Hageman G, Pachen D, Derom C, Vlietinck R, Zeegers MP. Placental telomere length decreases with gestational age and is influenced by parity: A study of third trimester liveborn twins. Placenta. 2014;**35**:791-796. DOI: 10.1016/j.placenta.2014.05.010

[39] Sukenik-Halevy R, Amiel A, Kidron D, Liberman M, Ganor-Paz Y, Biron-Shental T. Telomere homeostasis in trophoblast and in cord blood cells from pregnancies complicated with preeclampsia. American Journal of Obstetrics and Gynecology. 2016;**214**:283.e1-283.e7. DOI: 10.1016/j.ajog.2015.08.050

[40] Perrem K, Colgin LM, Neumann AA, Yeager TR, Reddel RR. Coexistence of alternative lengthening of telomeres and telomerase in hTERT-transfected GM847 cells. Molecular and Cellular Biology. 2001;**21**(12):3862-3875

[41] Biron-Shental T, Liberman M, Elbaz M, Laish I, Sharony R, Amiel A. Telomere homeostasis in placentas from pregnancies with uncontrolled diabetes. Placenta. 2016;**44**:13-18. DOI: 10.1016/j.placenta.2016.05.009

[42] Biron-Shental T, Sukenik-Halevy R, Naboani H, Liberman M, Kats R, Amiel A.

Telomeres are shorter in placentas from pregnancies with uncontrolled diabetes. Placenta. 2015;**36**:199-203. DOI: 10.1016/j.placenta.2014.11.011

[43] Biron-Shental T, Sukenik-Halevy R, Sharon Y, Goldberg-Bittman L, Kidron D, Fejgin MD, et al. Short telomeres may play a role in placental dysfunction in preeclampsia and intrauterine growth restriction. American Journal of Obstetrics and Gynecology. 2010;**202**:381.e1-381.e7. DOI: 10.1016/j.ajog.2010.01.036

[44] Sukenik-Halevy R, Fejgin M, Kidron D, Goldberg-Bittman L, Sharony R, Biron-Shental T, et al. Telomere aggregate formation in placenta specimens of pregnancies complicated with pre-eclampsia. Cancer Genetics and Cytogenetics. 2009;**195**:27-30. DOI: 10.1016/j.cancergencyto.2009.03.015

[45] Biron-Shental T, Sukenik-Halevy R, Goldberg-Bittman L, Kidron D, Fejgin MD, Amiel A. Telomeres are shorter in placental trophoblast of pregnancies complicated with intrauterine growth restriction (IUGR). Early Human Development. 2010;**86**:451-456. DOI: 10.1016/j.earlhumdev.2010.06.002

[46] Biron-Shental T, Kidron D, Sukenik-Halevy R, Goldberg-Bittman L, Sharony R, Fejgin MD, et al. TERC telomerase subunit gene copy number in placentas from pregnancies complicated with intrauterine growth restriction. Early Human Development. 2011;**87**:73-75. DOI: 10.1016/j.earlhumdev.2010.08.024

[47] Lehner R, Bobak J, Kim NW, Shroyer KR. Localization of telomerase hTERT protein and survivin in placenta: Relation to placental development and hydatiform mole. Obstetrics and Gynecology. 2001;**97**:965-970

[48] Biron-Shental T, Sukenik-Halevy R, Sharon Y, Laish I, Fejgin MD, Amiel A. Telomere shortening in intrauterine growth restriction placentas. Early Human Development. 2014;**90**:465-469. DOI: 10.1016/j.earlhumdev.2014.06.003

[49] Kudo T, Izutsu T, Sato T. Telomerase activity and apoptosis as indicators of ageing in placenta with and without intrauterine growth retardation. Placenta. 2000;**21**:493-500. DOI: 10.1053/plac.2000.0538

[50] Geifman-Holtzman O, Xiong Y, Holtzman EJ, Hoffman B, Gaughan J, Liebermann DA. Increased placental telomerase mRNA in hypertensive disorders of pregnancy. Hypertension in Pregnancy. 2010;**29**:434-445. DOI: 10.3109/10641950903214625

[51] Izutsu T, Izutsu N, Iwane A, Takada A, Nagasawa T, Kanasugi T, et al. Expression of human telomerase reverse transcriptase and correlation with telomerase activity in placentas with and without intrauterine growth retardation. Acta Obstetricia et Gynecologica Scandinavica. 2006;**85**:3-11

[52] Izutsu T, Kudo T, Sato T, Nishiya I, Ohyashiki K, Nakagawara K. Telomerase and proliferative activity in placenta from women with and without fetal growth restriction. Obstetrics and Gynecology. 1999;**93**:124-129

[53] Izutsu T, Kudo T, Sato T, Nishiya I, Ohyashiki K, Mori M, et al. Telomerase activity in human chorionic villi and placenta determined by TRAP and in situ TRAP assay. Placenta. 1998;**19**:613-618

[54] Moyzis RK, Buckingham JM, Crams LS, Dani M, Deaven LL, Jones M, et al. A highly conserved repetitive DNA sequence, $(TTAGGG)_n$, present at the telomeres of human chromosomes. Proceedings of the National Academy of Sciences. 1988;**85**:6622-6626

[55] Kimura M, Stone RC, Hunt SC, Skurnick J, Lu X, Cao X, et al. Measurement of telomere length by

the Southern blot analysis of terminal restriction fragment lengths. Nature Protocols. 2010;**5**:1596-1607. DOI: 10.1038/nprot.2010.124

[56] Cawthon RM. Telomere measurement by quantitative PCR. Nucleic Acids Research. 2002;**30**:e47

[57] Gil ME, Coetzer TL. Real-time quantitative PCR of telomere length. Molecular Biotechnology. 2004;**27**: 169-172. DOI: 10.1385/MB:27:2:169

[58] Cawthon RM. Telomere length measurement by a novel monochrome multiplex quantitative PCR method. Nucleic Acids Research. 2009;**37**:e21. DOI: 10.1093/nar/gkn1027

[59] Martin-Ruiz CM, Baird D, Roger L, Boukamp P, Krunic D, Cawthon R, et al. Reproducibility of telomere length assessment-an international collaborative study. International Journal of Epidemiology. 2015;**44**: 1749-1754. DOI: 10.1093/ije/dyv171

[60] O'Callaghan NJ, Fenech MA. Quantitative PCR method for measuring absolute telomere length. Biological Procedures Online. 2011;**13**:3. DOI: 10.1186/1480-9222-13-3

[61] Baird DM, Rowson J, Wynford-Thomas D, Kipling D. Extensive allelic variation and ultrashort telomeres in senescent human cells. Nature Genetics. 2003;**33**:203-207. DOI: 10.1038/ng1084

[62] Lansdorp PM, Verwoerd NP, Rijke FM, Dragowska V, Little MT, Dirks RW, et al. Heterogeneity in telomere length of human chromosomes. Human Molecular Genetics. 1996;**5**(5):685-691

[63] de Pauw ES, Verwoerd NP, Duinkerken N, Willemze R, Raap AK, Fibbe WE, et al. Assessment of telomere length in hematopoietic interphase cells using in situ hybridization and digital fluorescence microscopy. Cytometry. 1998;**32**:163-169

[64] Amiel A, Goldzak G, Gaber E, Yosef G, Fejgin MD, Yukla M, et al. Random aneuploidy and telomere capture in chronic lymphocytic leukemia and chronic myeloid leukemia patients. Cancer Genetics and Cytogenetics. 2005;**163**(1):12-16

[65] Morrish TA, Bekbolysnov D, Velliquette D, Morgan M, Ross B, Wang Y, et al. Multiple mechanisms contribute to telomere maintenance. Journal of Cancer Biology and Research. 2013;**1**:1012-1036

[66] Morrish TA, Greider CW. Short telomeres initiate telomere recombination in primary and tumor cells. PLoS Genetics. 2009;**5**:e1000357. DOI: 10.1371/journal.pgen.1000357

[67] Goodwin E, Meyne J. Strand-specific FISH reveals orientation of chromosome 18 alphoid DNA. Cytogenetics and Cell Genetics. 1993;**63**:126-127

[68] Goodwin EH, Meyne J, Bailey SM. Strand-specific in-situ hybridization reveals long-range molecular order in repetitive DNA. Cytogenetics and Cell Genetics. 1993;**63**:253

[69] Bailey SM, Goodwin EH, Cornforth MN. Strand-specific fluorescence in situ hybridization: The CO-FISH family. Cytogenetic and Genome Research. 2004;**107**(1-2):14-17. DOI: 10.1159/000079565

[70] Bailey SM, Brenneman MA, Goodwin EH. Frequent recombination in telomeric DNA may extend the proliferative life of telomerase-negative cells. Nucleic Acids Research. 2004 Jul 16;**32**(12):3743-3751

[71] McKenna MJ, Robinson E, Goodwin EH, Cornforth MN, Bailey

SM. Telomeres and NextGen CO-FISH: Directional genomic hybridization (Telo-dGH™). Methods in Molecular Biology. 2017;**1587**:103-112. DOI: 10.1007/978-1-4939-6892-3_10

[72] Hultdin M, Grönlund E, Norrback KF, Eriksson-Lindström E, Just T, Roos G. Telomere analysis by fluorescence in situ hybridization and flow cytometry. Nucleic Acids Research. 1998;**26**(16):3651-3656

[73] Baerlocher GM, Lansdorp PM. Telomere length measurements in leukocyte subsets by automated multicolor flow-FISH. Cytometry. Part A. 2003;**55**(1):1-6

[74] Baerlocher GM, Lansdorp PM. Telomere length measurements using fluorescence in situ hybridization and flow cytometry. Methods in Cell Biology. 2004;**75**:719-750

[75] Baerlocher GM, Vulto I, de Jong G, Lansdorp PM. Flow cytometry and FISH to measure the average length of telomeres (flow FISH). Nature Protocols. 2006;**1**:2365-2376

[76] Meeker AK, Gage WR, Hicks JL, Simon I, Coffman JR, Platz EA, et al. Telomere length assessment in human archival tissues: Combined telomere fluorescence in situ hybridization and immunostaining. The American Journal of Pathology. 2002;**160**(4):1259-1268

[77] Ding Z, Mangino M, Aviv A, Spector T, Durbin R, UK10K Consortium. Estimating telomere length from whole genome sequence data. Nucleic Acids Research. 2014;**42**:9. DOI: 10.1093/nar/gku181

[78] Mensa E, Latini S, Ramini D, Storci G, Bonafe M, Olivieri F. The telomere world and aging: Analytical challenges and future perspectives. Ageing Research Reviews. 2019;**50**:27-42. DOI: 10.1016/j.arr.2019.01.004

[79] Kim NW, Piatyszek MA, Prowse KR, Harley CB, West MD, Ho PL, et al. Specific association of human telomerase activity with immortal cells and cancer. Science. 1994;**266**(5193):2011-2015

[80] Feretzaki M, Lingner J. A practical qPCR approach to detect TERRA, the elusive telomeric repeat-containing RNA. Methods. 2017;**114**:39-45

[81] Porro A, Feuerhahn S, Delafontaine J, Riethman H, Rougemont J, Lingner J. Functional characterization of the TERRA transcriptome at damaged telomeres. Nature Communications. 2014;**31**(5):5379

[82] Sun D, Lopez-Guajardo CC, Quada J, Hurley LH, Von Hoff DD. Regulation of catalytic activity and processivity of human telomerase. Biochemistry. 1999;**38**(13):4037-4044

[83] Grolimund L, Aeby E, Hamelin R, Armand F, Chiappe D, Moniatte M, et al. A quantitative telomeric chromatin isolation protocol identifies different telomeric states. Nature Communications. 2013;**4**:2848

[84] Allsop R, Shimoda J, Easa D, Ward K. Long telomeres in the mature human placenta. Placenta. 2007;**28**:324-327. DOI: 10.1016/j.placenta.2006.04.003

[85] Rama S, Suresh Y, Rao AJ. Regulation of telomerase during human placental differentiation. A role of TGFbeta1. Molecular and Cellular Endocrinology. 2001;**182**:233-248

[86] Nishi H, Yahata N, Ohyashiki K, Isaka K, Shiraishi K, Ohyashiki JH, et al. Comparison of telomerase activity in normal chorionic villi to trophoblastic diseases. International Journal of Oncology. 1998;**12**:81-85

[87] Nishi H, Ohyashiki K, Fujito A, Yahata N, Ohyashiki JH, Isaka K, et al. Expression of telomerase subunits and

localization of telomerase activation in hydatiform mole. Placenta. 1999;**20**: 317-323. DOI: 10.1053/plac.1998.0386

[88] Nishi H, Nakada T, Kyo S, Inoue M, Shay JW, Isaka K. Hypoxia-inducible factor 1 mediates upregulation of telomerase (hTERT). Molecular and Cellular Biology. 2004;**24**:6076-6083. DOI: 10.1128/MCB.24.13.6076-6083.2004

[89] Carter AM. Placental oxygen consumption. Part I: In vivo studies-a review. Placenta. 2000;**21**(Suppl. A): S31-S37. DOI: 10.1053/plac.1999.0513

[90] Kuhlow D, Florian S, von Figura G, Weimer S, Schulz N, Petzke KJ, et al. Telomerase deficiency impairs glucose metabolis and insulin secretion. Aging. 2010;**2**:650-658

[91] Jirkovská M, Kučera T, Kaláb J, Jadrníček M, Niedobová V, Janáček J, et al. The branching pattern of villous capillaries and structural changes of placental terminal villi in type 1 diabetes mellitus. Placenta. 2012;**33**:343-351. DOI: 10.1016/j.placenta.2012.01.014

[92] Unek G, Ozmen A, Mendilcioglu I, Simsek M, Korgun ET. The expression of cell cycle related proteins PCNA, Ki67, p27 and p57 in normal and preeclamptic human placentas. Tissue & Cell. 2014;**46**:198-205. DOI: 10.1016/j.tice.2014.04.003

[93] Unek G, Ozmen A, Ozekinci M, Sakinci M, Korgun ET. Immunolocalization of cell cycle proteins (p57, p27, cyclin D3, PCNA and Ki67) in intrauterine growth retardation (IUGR) and normal human term placentas. Acta Histochemica. 2014;**116**:493-502. DOI: 10.1016/j.acthis.2013.10.007

[94] Ahmed W, Lingner J. Impact of oxidative stress on telomere biology. Differentiation. 2018;**99**:21-27. DOI: 10.1016/j.diff.2017.12.002

[95] Zinková A, Brynychová I, Svačina A, Jirkovská M, Korabečná M. Cell-free DNA from human plasma and serum differs in content of telomeric sequences and its ability to promote immune response. Scientific Reports. 2017;7:2591. DOI: 10.1038/s41598-017-02905-8